*f***P**

SHADOWS BRIGHT AS GLASS

*The Remarkable Story of One Man's Journey
from Brain Trauma to Artistic Triumph*

Amy Ellis Nutt

FREE PRESS

New York London Toronto Sydney

*f*P
Free Press
A Division of Simon & Schuster, Inc.
1230 Avenue of the Americas
New York, NY 10020

First Free Press hardcover edition April 2011

FREE PRESS and colophon are trademarks of Simon & Schuster, Inc.

For information about special discounts for bulk purchases,
please contact Simon & Schuster Special Sales at 1-866-506-1949 or
business@simonandschuster.com.

The Simon & Schuster Speakers Bureau can bring authors to your
live event. For more information or to book an event contact the
Simon & Schuster Speakers Bureau at 1-866-248-3049 or
visit our website at www.simonspeakers.com.

Book design by Ellen R. Sasahara

Lyrics from Bob Dylan's "Desolation Row" courtesy of Bob Dylan Music
Company. Copyright © 1965 by Warner Bros. Inc.; renewed 1993 by Special
Rider Music. International copyright secured. Reprinted by permission.

Manufactured in the United States of America

1   3   5   7   9   10   8   6   4   2

Library of Congress Cataloging-in-Publication Data

Nutt, Amy Ellis.
Shadows bright as glass: the remarkable story of one man's journey
from brain trauma to artistic triumph
/ Amy Ellis Nutt.
p. cm.
Includes bibliographical references and index.
1. Sarkin, Jon, 1953—Health. 2. Sarkin, Jon, 1953—Psychology. 3. Cerebrovas-
cular disease—Patients—United States—Biography. 4. Brain—Hemorrhage—
Patients—United States—Biography. 5. Artists—Psychology. I. Title. II. Title:
Accidental artist and the search for the soul.
RC388.5.N89 2011
362.196'810092—dc22
[B]   2010030162

ISBN 978-1-4391-4310-0
ISBN 978-1-4391-5007-8 (ebook)

For Jon and Kim

When one has weighed the sun in the balance, and measured the steps of the moon, and mapped out the seven heavens star by star, there still remains oneself. Who can calculate the orbit of his own soul?

—Oscar Wilde, from *De Profundis*

CONTENTS

SHADOWS BRIGHT AS GLASS

Prologue

Shadows Bright as Glass

The swirling waters along the North Shore of Boston are anchored by a geography of grief: Great Misery Island, Cripple Cove, the reef of Norman's Woe. Jon Sarkin is comfortable in this landscape, shaped as it is by loss. Centuries of women and children have waited on its rocky promontories for husbands and sons and fathers who never came home. Just west of Ten Pound Island, the Annisquam River empties into the Atlantic, watched over by the ghosts of Gloucester. In this colonial fishing village the sunlight still tastes of brine, and the oldest homes bear plaques inscribed with the names of Gloucestermen long dead: Colonel Joseph Foster, a veteran of the Revolution, who smuggled goods into Massachusetts during the British blockade of New England's harbors; Captain Harvey Coffin Mackay, whose sloop was struck by lightning and sank on its way to England in 1830; and the Luminist painter Fitz Hugh Lane, who immortalized that seafaring tragedy months later in his watercolor *The Burning of the Packet Ship* Boston.

For ages, artists have been summoned here by the views of ships' masts tangling in the harbor and Creamsicle-colored sunsets melting on the rocks. Winslow Homer visited and painted his *Boy on the*

Rocks. Rudyard Kipling vacationed and wrote *Captains Courageous,* and when Longfellow stopped for a look, he penned "The Wreck of the Hesperus."

> *And ever the fitful gusts between*
> *A sound came from the land;*
> *It was the sound of the trampling surf,*
> *On the rocks and hard sea-sand.*

A hundred years after Longfellow, T. S. Eliot remembered his childhood summers in Gloucester and wrote about the dangerous rocks hidden beneath the harbor's waves in his *Four Quartets.*

> *The river is within us, the*
> *sea is all about us;*
> *The sea is the land's edge also*

Art did not lure Jon Sarkin here, but it saved him. When he first arrived thirty years ago, he was a young, ambitious chiropractor intent on building a career. That was before his future slipped away from him, before a tiny blood vessel deep in his brain inexplicably shifted a hundredth of an inch, and as quickly as the flap of a butterfly's wing, set off a wave of events that altered him body and soul. A single cruel trick of nature, a catastrophic stroke, and a quiet, sensible man was transformed into an artist with a ferocious need to create.

For nearly two decades, he toiled in his studio painting and drawing without forethought or expectation, without plan or picture in his head, producing a storm of art that slowly increased in complexity and quality.

Yet always there was this question: Who was he? How had he gotten to this place? He was that rarest of individuals, a man dislocated from his own sense of self, a man who knew his brain had betrayed him and cast him out to sea. Recovered, it was as if he'd washed up on some alien shore, and he questioned who, and what, he was. How does a soul start over?

Nearly two thousand years ago, Plutarch asked the same question and was left puzzled. He wrote about a great ship that hardworking Athenians replaced plank by plank as the ship decayed until there was nothing left of the original. What was it now? Was it the same ship? Plutarch wondered. Or was it something wholly new?

Sarkin's body was broken, his brain dislocated. Parts of him were missing and others unrecognizably changed. He knew it, felt it deeply and yet could not explain how or why even to himself. To truly understand what had happened, he would have to be both subject and object, actor as well as audience. He was, in a way, his own philosophy experiment: How many pieces could be removed and replaced, without him becoming a different man?

This was a question the Ancients pondered, but neuroscientists now try to resolve as they search for the sources of consciousness. Sarkin, though, was an unwitting participant. Dislodged from himself, he had no choice but to find a way back in. He understood exquisitely, painfully, in a way few individuals can, that when the rock of his identity cracked, it let loose his own unsuspecting soul.

The patient blinked, wide-awake, as the surgeon peeled back the outer covering of the man's brain and began looking for the tumor.

"Soon, you'll be back on the ward," said one of the nurses in the operating room at the Cardiff, Wales, Royal Infirmary.

"Thank you, I feel fine," the man answered, his scalp numbed only by a local anesthetic.

The year was 1938 and the forty-one-year-old neurosurgeon, Lambert Rogers, had no MRI, not even a microscope, to locate his patient's brain tumor, just his probing fingers. After cutting through the transparent "skin" of the dura mater, the tough outer covering of the brain, Rogers plunged his hand into the soft, wet folds of the man's gray matter and began exploring. Half an hour went by. Then an hour.

The surgeon rummaged through the three-pound gelatinous mass like a blind man slogging through a swamp.

Christ, there are still two more patients to go, thought Wilfred Abse, the young intern assisting Rogers. Abse was a twenty-three-year-old psychiatrist-in-training, impatient to finish his long day. He also knew time was running out for the patient on the table. Abse had never seen a living human brain before, but as he watched Rogers poke and prod, each unsuccessful foray seemed to mutilate more of the poor man's brain tissue. Nearly two hours into the operation, his blood pressure dropped precipitously and he lost consciousness. If Rogers didn't find the tumor soon, he would have to sew the man back up. Still, he pressed on, the light from the small lamp strapped to his forehead illuminating the oozing pink and gray brain inside his patient's skull. What happened next Abse never forgot and often told his family about, as if still amazed after all the intervening years. The patient, who had been unresponsive for some time, suddenly cried out in a voice that seemed more mechanical than human:

"You sod, leave my soul alone. Leave . . . my . . . soul . . . alone."

Stunned, Rogers hastily removed his hand. It was as if the man's very spirit had reached out to defend itself. Within seconds, the beleaguered patient expired and a deep, disturbed silence descended on the room.

Sarkin's soul had been abused, too, his brain laid bare and unraveled nearly to its core. To find a way back in, to rewrite the narrative of his self, he would have to weave together all the threads, old and new. We are, each of us, a story, a self born of a billion babbling neurons. Liquid possibility. In such fragile seas, how does a single soul stay afloat?

Chapter 1

THIN AS A THREAD

There was money riding on the hole, as always. The two friends had battled back and forth all afternoon on the Cape Ann golf course, and though Hank Turgeon was finally ahead by a couple of bucks, he wasn't about to let up. He'd already sprung for the beer: twelve mini cans of Budweiser stashed in ice in the side pocket of his golf bag. The beer probably weighed more than his clubs did—he carried just three irons and a driver. Jon Sarkin, however, had about a dozen irons and woods in his father's old canvas bag. The competition may have been friendly, but Sarkin liked to be prepared for contingencies—the rough, trees, water—and Lord knows there were plenty of them when he played. He needed all the clubs he could carry.

By the time they reached the eighth hole, they were loose and laughing, enjoying their good-natured rivalry. Turgeon had the honor, stepped up to the tee, and took a healthy hack with his driver. He hit the ball fat and popped it high into the crisp, Wedgwood-blue sky, then watched it land in the thick rough, just fifty yards away.

"Nice shot," Sarkin said sarcastically.

The time was about 3:00 p.m., Thursday, October 20, 1988, and the two friends had lucked out when they decided to cut out of work early, Sarkin from his chiropractic office, Turgeon from his carpentry. The weather was warm and sunny, and they both happily breathed in the ocean air wafting across the Cape Ann golf course, some thirty-five miles northeast of Boston. The nine public holes rest on a spit of land carved out of the swamp and rocks by glaciers thousands of years ago. Sarkin always enjoyed the view from the eighth tee box, the highest point on the course and the farthest from the clubhouse. To his left, a winding creek emptied into a thin slice of harbor that gave way to the yawning Atlantic. To his right, shards of sunlight splintered the last leaves on a tall oak, scattering shadows across the fairway. A slight breeze rippled the surface of the creek as Sarkin bent down, reached inside the pocket of his golf bag, and fished around for a tee. As he pulled his hand out, he experienced a hideous dizzying sensation, as if his brain had suddenly twisted inside his head. He stood up and froze.

What the hell just happened? he thought.

In less than half a second, a part of his head had seemed to unhinge, to split apart and rush away.

I'm going to die. I'm thirty-five years old and I'm going to die, he said to himself.

"Is anything wrong?" Turgeon asked.

Sarkin hesitated, trying to get his bearings.

"Nope."

What *could* he say? That he felt as if his brain had just broken in half? Maybe the sensation would pass. Maybe he even imagined it. He took a few deep breaths, teed up his ball, and swung from his heels. As he often did on his drives, he topped the ball twenty yards—plunk—right into the marsh at the front of the tee box.

"You're going to break your freakin' neck with that swing." Turgeon laughed.

Sarkin felt queasy, and as he walked toward the fairway, he tried not to move his head. What he did not know—what he could not know—was that somewhere deep in his brain, a single blood vessel had shifted ever so slightly and the movement, as miniscule as it was, had caused a cataclysmic response in one of his cranial nerves.

There are 100 billion capillaries in the human brain. Placed end to end they would stretch from Philadelphia, Pennsylvania, to Portland, Maine. Inside Sarkin's head a tiny patch of one of those blood vessels, as narrow as a thread and no longer than a single stitch, had suddenly bulged and was now touching, ever so slightly, the eighth cranial nerve. In that thinnest of breaths between one moment and the next, Sarkin's hearing and balance were threatened and, if the vessel ruptured, possibly his life. He felt dizzy and nauseous and confused.

Sarkin didn't bother looking for his ball, so he took a drop. Several strokes later, he finally chopped and hacked his way to the green. All he wanted was to go home.

"Do you mind if we quit?" he said after putting out.

"Sure," Turgeon answered.

Sarkin had become quiet and Turgeon thought that maybe his friend was just frustrated with his game. No problem. He'd had enough, too, and it was getting cool, anyway. In the fading autumn light, they wound their way back to the clubhouse, dragging their clubs behind them.

On the ten-minute drive home to Gloucester, Sarkin sat glumly in the passenger seat of Turgeon's navy-blue pickup truck, trying to right himself. Looking out the window at the autumn colors whirling by, he started to feel dizzy again. A sense of panic, even dread, enveloped him. He had no idea what was going to happen next. Often, after they played a round, the two friends drove up to Halibut Point for drinks, but Turgeon sensed Sarkin just wanted to get home.

The two had known each other since the early 1980s. Both played guitar and both occasionally jammed with a local band called the Joe Tones. Turgeon grew up on Cape Ann, and when he first met Sarkin, he wasn't sure what to think. Here was this Eastern college boy in khaki pants and pressed shirt, smart and polite and straight-edged. But the guy sure knew music—jazz, folk, blues, rock. In the past several years, they had also become dedicated duffers. They played at least once a week after work or on the weekends. One summer they even rented a house in Truro, out on Cape Cod, and played thirty-six holes a day for a week. They were out in the sun so long they told each other they looked like burn victims.

By the time Turgeon dropped Sarkin off it was almost dark. He got out of the truck and slowly lifted his clubs from the bed.

"See ya," he said.

Sarkin paced himself, walking carefully up the curving granite steps to the front door, still hoping the world might tilt back on its axis. When he walked in the door, his wife, Kim, knew immediately something wasn't right. He looked miserable.

"What's wrong, Jon?" she asked, balancing their nine-month-old baby boy on her hip. She watched as her husband walked slowly across the living room to the futon couch, sat down, and buried his head in his hands.

"What's wrong?" she asked again, a little more urgently.

"I don't know how to explain it," he answered "Something happened. I was bending down, and then my brain just . . . twisted."

Sarkin held two clenched fists in front of his face and then turned them, abruptly, in opposite directions, as if wringing out a wet towel.

"I don't know what happened," he said. "I just know everything is different. Everything's different and it's not ever going to be the same."

Chapter 2

A Thousand Screaming Baboons

Nothing about the human brain is without paradox. Cocooned in its mother's womb, a fetus at six weeks has the nervous system of a shrimp, and yet by the time children reach puberty their brains have become the most complex objects in the universe, tightly wound balls of nerve fibers, which, if placed end to end, would stretch for 100,000 miles—enough to circle the globe four times.

A three-pound mass of tissue and fluid inspired the music in Mozart's fingers, the poetry in Shakespeare's words and the mathematics of Newton's imagination. And yet this vaunted cathedral of thinking weighs no larger than the average-sized turnip, transmits signals three million times slower than a wire does electricity, and runs on a mere 20 watts of energy (just enough to power the light in your refrigerator). Scientists still know only a fraction about the brain, even less about why it sometimes doesn't work.

Three days after experiencing the strange shudder deep inside his head, Sarkin began hearing a high-pitched screech. The sound grew louder and shriller every day until, by the end of October, it was nearly deafening.

"Like a thousand screaming baboons," he told Kim.

Each morning when he opened his eyes, he prayed the excruciating noise would be gone, and when it wasn't, he forced himself to soldier on. He rose at 8:00 a.m. and dressed in his usual khaki pants, white shirt with tie, and blue blazer. With difficulty he bent down to put on his shoes, a pair of well-worn wingtips. He ate little for breakfast, too distracted by the sounds in his head, but every day he drove to work, a chiropractic office in nearby Hamilton. After the incident on the golf course, he no longer turned on the radio on his way into work. How could he listen to rock 'n' roll with this screaming in his head? He told no one except Kim about what he was going through, mainly because he felt helpless to explain something *he* barely understood. For two weeks Sarkin tried to maintain a normal schedule, seeing a half dozen patients, manipulating their necks and backs and jawbones, as if nothing had changed in his life. But in the quiet of his office, as he wrote up his notes each afternoon, he would dim the lights, fold his arms on his desk, and lay his head down in despair.

In November, he went to see his friend and doctor, John Abramson, who gave him a thorough neurological exam. He tested Sarkin's balance, the reflexes in the tendons of his arms and legs, his vision, facial sensation, and the response of his pupils to light. He tapped Sarkin's chin to check for "jaw jerk," a sign of a brain lesion, and asked him to stick out his tongue to see if it flicked back and forth, a condition known as "trombone tremor" and a possible indication of a movement disorder. Then Abramson extended his hands toward Sarkin and asked him to grab both and hold tight. If Sarkin's grasp quickly slackened, Abramson would know he had "milkmaid's grip," just one symptom of many that could point to a neurodegenerative disease. He also probed Sarkin's ears for obstructions, fluid, or damage and gave him a basic hearing test, placing a tuning fork on the mastoid bone behind each ear.

Everything came up normal.

Abramson didn't doubt Sarkin was suffering—that much was clear from his friend's haggard look and the way he held his body, rigid and controlled, as if in a constant cringe. All the doctor could tell him, though, was that he probably had a condition known as tinnitus (Latin for "ringing"), where the sound comes not from outside, but rather inside the head. For most tinnitus patients, the problem arises from hearing loss due to age, trauma, exposure to loud noises, or even drugs, all of which can damage the cochlea, the part of the ear that allows us to hear.

The earliest scientific records that refer to a tinnitus-like condition can be traced to the ancient Babylonians, who described "singing ear" and "whispering ear," for which the treatment was usually an incantation to drive out evil spirits. Today specialists know that tinnitus is physiological in origin and can occur to varying degrees from mild to severe in millions of people. Doctors don't, however, know how or why tinnitus occurs. The most popular idea is that the brain is confused, unable to figure out the abnormal signals being sent to it from the damaged cochlea, so it simply tries to "fill in" the missing sounds with noise of its own.

For a much smaller number of tinnitus sufferers, the cause of the problem is not the cochlea, but rather the brain and a disturbance of the auditory nerve. For this type of tinnitus the only treatments available are sedatives. All Abramson could do was categorize Sarkin's clinical symptoms, which for tinnitus are rated on a sliding scale, beginning at the bottom with "slight." Sarkin's symptoms topped out at "catastrophic."

By December, the situation had worsened. Not only was the noise in his head deafening, but also every other sound was now causing him extreme discomfort. About a third of those who suffer from tinnitus also experience normal sounds as loud and abrasive, a condition known as hyperacusis. Sometimes this sensory sensitivity spills over into an intolerance for bright lights, strong smells, and motion, all of which Sarkin now experienced. His senses were muddled and confused, but sound was particularly torturous and threatened him everywhere: a fork scraping across a plate, someone chewing, Kim turning the pages of the newspaper. Even when Sarkin lay in bed in the morning, the crackling coming from the kitchen sounded less like eggs frying than gunfire. Each burst blistered his ears and sent him digging deeper under the bedcovers.

Kim whispered now more than she talked and avoided conversations on the phone when Jon was home. She went shoeless, too, practically tiptoeing across the hardwood floors, and even eviscerated the baby's toys, removing the tinkling bells and whistles that could send Jon into paroxysms of agony. Early one evening, after dragging himself to work and back home again, Sarkin pulled slowly into the driveway. The racket coming from his car crunching across the gravel made him wince. He got out and barely clicked the door closed behind him, then climbed the steps to the house. He had laid these cobblestones himself, just a few months earlier. When he opened the front door, Kim was in the kitchen, at the other end of the hallway, and called out to her husband.

"Hi, hon, how are you?"

This is what she always asked when Jon came home, but since October the question was freighted with anxiety. Just inside the door, Sarkin placed his briefcase on the ground, leaned back against the wall, and slowly slid down to the floor. Curled up in a fetal position, he began to cry. For a moment, Kim stood there, in shock. She had never seen her husband cry, not once in the five years they'd known each other. She lowered herself until she was sitting next to him, touched him gently, and tried to comfort him.

"My life is hell," he said to Kim.

"Mine, too," she answered.

Chapter 3

ROLLING THE DICE

Three chiropractors, a neurologist, a psychiatrist, even an acupuncturist—Sarkin tried all of them during the winter months of 1988 and throughout 1989. He also tried to maintain his office schedule, although on some days he worked shorter hours and on others canceled appointments. When he wasn't at his office, he was at the library reading medical textbooks, scouring them for information about tinnitus and hyperacusis. He spoke less and less at home, and spent more time in the bedroom, with the lights turned off and the covers pulled over his head. Desperate for help at one point, he called a local tinnitus support group.

"Sure, we'd love to have you," the man on the other end said. Then he asked, "Did you ever notice the tinnitus getting worse with the full moon?"

Sarkin never went to a meeting. More depressed than ever, he sometimes found himself wishing the sun would not come up the next day. Kim tried to be supportive, mostly by making the house as calm, quiet, and comfortable as possible. She let him withdraw when he needed to, but when his symptoms failed to subside after so many months, she began to feel helpless—and worse, hopeless. A friend suggested one day that maybe Jon should be in a mental hospital.

"That wouldn't help," she told her blankly.

She never doubted the severity of Jon's condition, but she was upset that his illness was pulling him away from the family. Luckily, she had the baby, Curtis, and she poured herself into caring for him, cherishing being a mother. Before the explosion in his head, which is how Sarkin had come to describe what happened to him on the golf course, life with Kim had been blissful, and though he was now riddled with pain and anxiety almost every minute of the day, he still found solace in sleep, especially with Kim and the baby beside him. They had been married only two years when Curtis was born. Getting pregnant had been difficult—it took nine months—but both Kim and Jon were determined to have a family. She had always told him she wanted at least three children. That was the minimum, she said, and Jon had happily agreed. Now their plans for more kids were on hold.

By the beginning of the summer Jon, too, was beginning to feel hopeless about his situation. At dinner one night, as Curtis played in his high chair, Sarkin blurted out, "I don't think we can have more kids."

For a few seconds, Kim was taken aback.

"Jon, you don't really mean that. It's just now; it's just because of what you're going through. You won't always be this way."

The idea of not having more babies was something she couldn't even think about, much less talk about. She quickly changed the subject.

Some in Sarkin's family urged him to see a psychiatrist, hoping that the tinnitus might be more psychological than physiological. Sarkin didn't think so, of course, but he didn't want to dismiss any possible treatment, either. He visited a psychiatrist with a Freudian bent who was intrigued by the fact that Sarkin's father had enjoyed playing golf and that his patient's "cerebral event" had occurred on a golf course. Sarkin was not impressed. He stayed in treatment only long enough to try antidepressant medication, and when that didn't work, painkillers, sedatives, even an anticonvulsant normally given to patients with epilepsy. Valium brought a modicum of relief, but it always seemed to take too long and lasted even shorter. Finding a way out of his desolation was the first thing Sarkin thought about when he woke up each morning and the last thing on his mind before falling asleep.

Sometime that summer, a friend from town who taught meditation stopped by, lingering until dinnertime. Jon and Kim invited him to stay to eat. That was unusual, because neither of them had told many people beyond their families about what had happened to Jon. It was just too hard to explain. There was no wound to show, no broken bone. He was tired and drawn, but to most people who didn't know better, everything still seemed normal. How bad could it be?

Trying to keep up appearances, even talking for an extended period of time, was difficult for Sarkin. It took all of his strength just to participate in the conversation that night, and before the dinner was over, Sarkin excused himself from the table. When he closed the door of the bathroom, he dropped to his knees, then all the way to the floor, and pressed his cheek against the cool linoleum. For a brief moment, it soothed him, but the wailing siren in his head started up again. Slowly Sarkin lifted himself off the floor and reached into the medicine cabinet for the bottle of Valium. It would take at least a half hour for the pill to start working. He couldn't wait that long, so he shook a single yellow tablet out of the bottle, rummaged around in a drawer until he found a nail clipper, then fell back onto the floor. Lying on his stomach, his face just inches above the linoleum, he used the blunt edge of the clipper to crush the Valium into a fine white powder, then lowered his face and snorted. For twenty, maybe forty, minutes he felt almost normal again.

When Sarkin's brother, Richard, a pediatrician in Buffalo, came to visit that summer for a couple of days, Jon was more animated than Kim had seen him in months. He even smiled and joked. She couldn't believe it, and when Richard left, she cornered Jon. She was frustrated, even a little angry.

"If you can act happy and well for Richard, then why can't you do that for me?" she challenged him.

He tried to explain that putting on the "show" for Richard had been exhausting; it took everything out of him. He certainly couldn't do it all the time. He didn't want his mother or brother or sister to worry about him. Kim already knew how bad it was, he said. She was the *only* one who knew how bad it was.

Not long afterward, at a weekly playgroup with other mothers and children, Kim finally broke down. Over the weeks and months she had begun to pour out her worries and frustrations to just a few of her friends, but she hadn't cried once—until now. For so long she'd held out hope that things would somehow get better with time, that Jon would either find a doctor with a miracle cure or he would just wake up and be normal again. But none of that had happened. In fact, in all these months, he hadn't improved at all.

My life as I know it is over, she said to herself. *It's over.*

Then, in July, Sarkin read an article about a tinnitus specialist in New York and immediately made an appointment for a few weeks later. For two days, the specialist put him through a battery of exams, brain scans, and hearing tests. The only way to measure tinnitus quantitatively, he explained, was to compare it to other sounds of known decibel levels. Bombarded with more than one sound at the same time, the human brain tends to pick out only the loudest. For the test, the doctor asked Sarkin to wear a pair of headphones and then played for him a series of increasingly loud notes, each time asking him whether he was still aware of the tinnitus. The higher the amplitude, or decibel level, of the notes at which the tinnitus could still be heard, the worse his condition. A ticking watch is about 20 decibels; an alarm clock, 80. The buzz from a power mower or chainsaw will measure around 105 decibels, rock concerts and thunderclaps, 120. At 130 decibels, the approximate loudness of a jackhammer or a jet engine, the sensation of sound is converted to pain.

At 130 decibels Sarkin could still hear the howling inside his head.

At the end of the second day, he sat across from the doctor and waited for his recommendation.

"You should go out and buy a white-noise generator."

Sarkin was stunned.

"That's it?" he asked, trying to control his bitter disappointment. "After all this, that's it? That's all you can suggest?"

He might as well have been back in the nineteenth century, or even earlier. The ancient Greeks and Romans prescribed everything from cucumber juice to ox urine for tinnitus. During the Middle Ages, some Englishmen cut a piping hot loaf of bread in half and placed each end against an ear to "sweat" the sound out. As far as masking the noise with a louder sound, the specialist's advice was more than two thousand years late. Pressed, however, the doctor offered one tantalizing long shot. Perhaps Sarkin's problem wasn't with his ear, but his brain. Something might be stressing his eighth cranial, or vestibulocochlear, nerve, which is responsible for hearing and balance. The eighth is a delicate thread of neurons that stretches from the medulla oblongata atop the spinal column, through the skull at the temporal bone and into the ear. If this was the source of his problem, said the doctor, then there was a neurosurgeon at Presbyterian University Hospital in Pittsburgh named Peter Jannetta whom Sarkin might want to consider. Jannetta had pioneered a specific procedure for cranial nerve disorders, but it involved a deep brain operation and the procedure wasn't yet fully accepted by the medical community.

The next day Sarkin flew home. For a while he sat in his car baking in the airport parking lot and watched the heat rise, wave after wave, from the blistering black pavement. He couldn't believe it. This was what it all came down to: either buy a white-noise machine or let a surgeon drill a hole in his head. He pounded the steering wheel again and again in anger and despair.

When he got home, he talked to Kim about Jannetta. At first she was horrified.

"Brain surgery, Jon? You want to have brain surgery?!"

She knew he was desperate, that he had tried everything, but letting someone cut into his brain? She worried just what an operation on his head would do to him, even if it worked. And what if it didn't? His mother, brother, and sister—even Abramson, the family doctor—

all tried to talk him out of it. Sarkin pressed on, though. He researched the procedure and found out it was both dangerous and controversial. The surgeon not only drilled a hole through the skull but also had to dig deep into the brain to isolate the problem along the auditory nerve. There was also no evidence that the operation would cure him. But at least it was something, he said to Kim.

In July 1989, Jon and Kim flew to Pittsburgh. Jannetta ran tests, scanned Sarkin's brain, and read through his medical records. Finally, he was ready to talk.

"It's pretty obvious to me what's going on," he told them.

A blood vessel had shifted deep inside Sarkin's brain and become swollen, he explained. Now it was impinging on his acoustic nerve, causing the strange and painful distortions of sound. Blood vessels sometimes moved during the aging process, he said. They can harden and lengthen, and when a blood vessel presses against a nerve in a certain way, and in a certain place, the nerve is unnaturally stimulated, causing a wide range of symptoms from extreme facial spasms and pain to hearing loss or tinnitus.

Jannetta walked them through the operation step by step, stopping every now and then to draw a diagram on a sheet of paper. He showed them where he believed the problem was, along the eighth cranial nerve, and described what he would do. Neither Jon nor Kim could believe it. He was not only telling them he knew the cause of Jon's suffering; he was saying he could cure it. Not just muffle the noise in his head, but eliminate it.

The operation, Jannetta explained, was extremely delicate. He would have to lift the troublesome blood vessel off the nerve and insert between the two a wafer-thin piece of Teflon fabric that would act as a kind of protective cushion. The dangers were myriad, he said—bleeding, seizures, stroke, even death—but he had done this operation hundreds of times and no one had ever died or even suffered a serious complication. Ninety-eight percent of his patients reported at least some relief of their symptoms and more than 80 percent of them complete relief. For those people like Sarkin who had suffered from tinnitus for two years or less, the success rate was 90 percent.

At the end of the consultation Sarkin told the surgeon that he needed to think about it, but he had already made up his mind. He was barely able to work, and when he came home, more often than not he immediately went to bed. Kim was effectively a single parent. Jon could barely hold Curtis unless the baby was sleeping. He had to undergo the surgery, he said to Kim. He had to try something. Kim worried that this was too big a step, and persuaded him to visit one more doctor, in Boston. The neurologist put Sarkin through an MRI, then showed him the scans.

"Do you see this blood vessel here?" he asked, pointing to a barely discernible shadow squiggling across the film. He had his finger on the same area Jannetta had highlighted.

"That might be the problem."

The neurologist insisted, however, that there was no way to really know. Dr. Jannetta, he explained, was a controversial figure because there was no concrete proof that compression of a nerve by a blood vessel could cause the kind of symptoms Sarkin described, especially when those symptoms were common to other conditions, such as Ménière's disease. Logically, there was nothing about Jon's tinnitus that pointed to a vascular cause. Even with the success rate of the operation, the objections of other surgeons were considerable, based on the potential for a catastrophic outcome. This particular operation carried a 1 percent mortality rate and a 10 percent chance of significant complications.

"I don't recommend this," said the neurologist.

On the forty-five-minute drive back to Gloucester, Jon and Kim had a serious talk.

"Do you really want to do this?" Kim finally asked.

"I've got to," Jon said.

Then he told her, for the first time, how often he thought about dying, that he had sometimes contemplated suicide.

"You can't do that, Jon. Promise me you won't ever do that," said Kim.

He felt guilty for bringing it up and told her he didn't really want to take his own life; it's just that he was so tired of it all he didn't want to wake up in the morning. When they got back home, Sarkin called his mother, brother, and sister and told them he was going ahead with the surgery. Then he called his doctor and friend, Abramson.

"Well," said the physician, "I guess you decided to roll the dice."

A few days before flying to Pittsburgh, Sarkin threw a going-away poker party at his house. Two of his best friends were there, his golf partner Hank Turgeon and John Keegan, a singer and songwriter who sometimes let Sarkin jam with Keegan's rhythm and blues band, Madhouse. Only an average poker player, Sarkin, more often than not, was on the losing side of their games, which also meant he was usually the guy forking over most of the cash: thirty bucks, forty, sometimes even a hundred. But on this night it was like someone had hit Sarkin's "boom switch." He was dealt one hot hand after another. Round after round he raked in the chips and watched his stacks rise and everyone else's diminish.

"The gods of poker are lookin' out for me, boys!" he cackled.

At the end of the night, Keegan alone owed Sarkin $500.

"I know you're good for it," he said to his friend.

If Sarkin needed an omen about his upcoming surgery, this surely was it. How could anything go wrong now?

Chapter 4

Six Cups of Fluid and Tissue

On August 8, 1989, only a few weeks after Sarkin and his wife visited Dr. Jannetta in Pittsburgh, the neurosurgeon fingered the bony knob just behind his patient's left ear and marked the shaved area with a large black "X." Sarkin was asleep on an operating table, his head secured in a surgical vise, his brain about to be exposed to the world. Several floors away his mother, Elaine, and Kim sat in a waiting room, bouncing the baby back and forth between them. Jannetta stopped to talk to them before scrubbing in. He wanted to let them know, one last time, what the possible complications could be: heart attack, seizure, stroke, coma, death.

Elaine jumped up and said, "Let's go."

She was joking, of course, but Kim also knew how scared she was. Jon's father died at age forty-nine of a heart attack in his sleep, so Elaine understood all too well how suddenly and unexpectedly someone you love can be taken from you. As for Kim, she had passed beyond the worry of the operation itself. Every surgeon had to say those things about stroke, coma, and death—that didn't mean they would actually happen. Jon had become so convinced the surgery would work she felt she had to back him, and now all she wanted to think about was whether the operation would cure him.

Jannetta's quarry, the source of Sarkin's symptoms, was a swollen blood vessel that could be anywhere along the length of the eighth cranial nerve. In order to find it, Jannetta would have to reach deep inside Sarkin's skull. Brain tissue is divided into two types: gray, which consists of the cell bodies of neurons, and white, the long filaments that connect those cell bodies and transmit electrical signals from one area of the brain to another. Some studies have shown a positive correlation between the density of gray matter, especially in the orbitofrontal cortex, and general intellectual ability. White matter, which gets its name from the fatty myelin sheets that encase the nerve fibers between neurons, governs the speed with which messages are sent between neurons. Disorders of white matter have been implicated in everything from multiple sclerosis to autism.

The trickiest part for Jannetta was actually not the journey through the gray and white matter. The hazard was the tangle of blood vessels and nerves at the bottom of the brain that crisscross a tiny space called the cerebellopontine angle. Bounded by the cerebellum and part of the temporal bone, the cerebellopontine angle is the gateway to the brain stem. Three major arteries and several nerves, including the vestibulocochlear, traverse the angle. This is where the eighth cranial nerve emerges from the brain stem, and so this is where Jannetta would have to begin his search. Damage virtually any other part of the brain, and the patient can recover function. Damage the brain stem, which controls the most basic elements of life—breathing, blood pressure, consciousness—and the result is usually death.

Sitting on a stool, with Sarkin's head bolted in place and practically in Jannetta's lap, the surgeon made a two-inch-long incision in his patient's scalp. Despite the doctor's beefy hands, his fingers moved nimbly and confidently, as if working the delicate strings of a marionette. He had been performing this type of surgery for twenty-two years, but only a few of these operations had involved the eighth cranial nerve. Mostly he employed the procedure on patients with facial spasms and neuralgia involving the fifth and seventh nerves.

Digging down through the skull, a surgeon is like a paleontologist tunneling back through time. Each part of the brain is a signpost to our evolutionary development, which began 500 million years ago with one of our most distant ancestors, fish. What set these creatures apart from sponges, jellyfish, and clams was a hardened tube of tissue called a backbone. The brains of all animals, including humans, can be traced back to the first spinal cord that ran down through the vertebrae of the first backbone. After our fish fathers crawled onto land, more brain matter was added on top of the spinal cord and more functions, from breathing, crawling, and walking to thinking, feeling, and remembering.

The outermost shell of the human brain contains six layers of neocortex less than a quarter-inch thick, formed just four million years ago. From these winding folds of gray matter, language, social behavior, and reasoning emerged. Beneath the neocortex is the 150-million-year-old cerebrum and the two hemispheres of the brain that were originally nothing more than a pair of small, smooth swellings, chiefly involved in olfaction, or the sense of smell. Next, at 300 million years old, are the mid-, or mammalian, brain and the limbic system, lying just under the cortex. The limbic system consists of several structures, including the amygdala and hippocampus, which are the primary processing centers for the emotions and memory, respectively. Deeper still is the 500-million-year-old hind-, or reptilian, brain, perched atop the summit of the spine and connecting all brain matter to the spinal cord.

Jannetta would have to reach through, and around, all three "brains" to find the troublesome blood vessel, and to make the final push to the stem he would have to delicately move aside the cerebellum, a curved lump of tissue about the size of a man's fist. The cerebellum is one of the least understood structures of the brain. Primarily it controls balance and the movement of muscles, but in recent years researchers have amassed evidence that damage to the cerebellum can result not only in motor dysfunction, but in a number of cognitive and behavioral problems as well, including personality changes, mood disorders, and attention deficit.

After slicing through Sarkin's scalp, Jannetta drilled into his quarter-inch-thick skull, clipping away pieces of bone until the rectangular opening to the brain was about the width of a quarter. Next, he cut through the tough, transparent outer covering of the brain and its last barrier of protection, the dura mater. He now had a clear view into Sarkin's brain—six cups of fluid and tissue containing 100 billion neurons and 100 trillion synapses, the way stations where information is passed from one neuron to the next. Every sensation, every idea, every action creates a unique firing pattern in the brain, and each firing pattern creates a new wave of activity that is constantly altering the tides of consciousness. Some scientists have suggested that there are more synapses in the human brain than there are atomic particles in the universe.

For the next stage of the operation, Jannetta would need the standing microscope in order to view the miniscule structures of the brain. From here on in, even a slight deviation from the path through to the back of Sarkin's brain could cause catastrophic damage, rendering him unable to hear, speak, or walk. The team worked quickly and efficiently, with the scrub nurse periodically flushing the brain and suctioning away excess blood as Jannetta cut through and cauterized minor capillaries obscuring his view.

Many surgeons, when not in the OR, eschew anything that might make their hands unsteady. Jannetta only recently had quit smoking and he still drank multiple cups of coffee a day. Often, if all was going well, he hummed during his operations. On this day, everything was running smoothly. Jannetta's work was so fine and focused he seemed not to be moving at all.

Using retractors, he carefully pulled aside Sarkin's left temporal lobe, where speech and facial recognition originate and long-term memories are stored. Now he could see the tentorium, a fibrous membrane that divides the upper and lower parts of the brain and protects the cerebellum. After cutting through the tentorium, Jannetta would need to move aside the cerebellum—"turning the corner," neurosurgeons call it—in order to reach the cerebellopontine angle and the eighth cranial nerve. The cerebellum sits at the very back of the skull and takes up only about 10 percent of the brain's volume but contains more than half of the brain's neurons. Many of those neurons contribute to the unconscious timing and coordination of muscles. Without them, it would be difficult to walk, talk, feed oneself, or even see straight. To slide around the cerebellum, Jannetta placed a piece of rubber from a surgical glove over cotton to form a kind of protective barrier so the cerebellum wouldn't be injured by the retractor holding it back.

Like a detective searching for a criminal, Jannetta followed Sarkin's ninth and tenth cranial nerves down to the brain stem. Along the way he encountered a small amount of bleeding near where the cerebrospinal fluid is produced, but the blood quickly coagulated and he moved on. Allowing any blood to build up could be catastrophic. Blood irritates brain tissue, causing it to swell, and both that swelling and the pooling blood can put pressure on the surrounding tissue, quickly destroying it. This deep in the brain, such damage could easily send a patient into a coma.

Jannetta had barely stirred in his seat for the past hour. He held his body and head motionless while he gently probed. Finally, he reached the eighth cranial nerve. Now he could clearly see the source of Sarkin's painful condition. In fact, there were two problems. First, just as he had predicted, a vein no more than four-hundredths of an inch thick (about the width of a single raindrop) was rubbing up against the auditory nerve. Jannetta carefully cut it away and cauterized the two ends. The second problem was more difficult and more perilous. The anterior inferior cerebellar artery, one of a pair of blood vessels that provide oxygen to portions of the frontal and parietal lobes, was twisted across the auditory nerve. If a surgeon nicks this artery, a stroke or heart attack can result and the damage can range from total deafness to "locked-in" syndrome, in which a patient is completely immobilized except for the ability to blink.

While Jannetta carefully lifted the artery off the nerve, the scrub nurse tore a quarter-inch piece of material from a nearby pad of Teflon, dipped it into a saline solution, and then, cradling it in the palm of her hand, rolled it into the shape of a tiny cigar. After placing the white fiber in a pair of micro-forceps, she handed the instrument to Jannetta, who cautiously placed the shredded piece of Teflon between the vessel and the nerve.

The main work was done. Now the surgeon needed to retrace his steps carefully through Sarkin's brain, making sure there was no bleeding and removing the retractor holding the cerebellum out of the way. Jannetta stitched the dura mater closed and inserted a rigid piece of mesh, called a cranioplasty plate, over the opening in Sarkin's skull. Finally, he let a resident sew up the incision in the scalp.

As Sarkin was wheeled into recovery, the surgeon greeted Kim and Elaine in the waiting room. The two had found an outlet for their anxiety by taking turns pushing Curtis around the hospital halls in his stroller. Everything had gone exactly as planned, Jannetta said, no complications. Kim could see her husband in about an hour.

Later, in the recovery room, she stood beside Jon's bed as he slowly woke. Bandages were wrapped around his head and his face was drained of color, but Kim wasn't concerned about that. She knew he would bounce back physically in no time. She was worried about only one thing and she needed Jon to tell her now. When he opened his eyes, she was looking at him.

"Jon, can you hear me? Is the ringing gone?"

He couldn't talk yet, but he understood, and he knew the answer. He nodded slowly, mouthing.

"Yes."

The nightmare was finally over. Ten months of anxiety, pain, and confusion slipped away with that one word. "Yes."

Kim had her husband back.

Chapter 5

BUILDING A LIFE

S arkin had taken a circuitous route before his marriage to Kim. It was in their relationship and in starting a family that he found his life's purpose and resolved previously competing impulses.

His early life had been full of contradictions. He was a kid who had both an artistic, unconventional side and a polite, play-by-the-rules side, mindful never to get into too much trouble. Growing up, his relationships with his parents and two siblings were close, and now, like his father, he had become a devoted family man. But finding the life he wanted had been a journey.

His earliest memory was his fourth birthday—not the actual celebration, but the warmth of the sunny April morning as he helped his mother tie balloons to the metal railing of the steps leading down to the backyard. Color, shape, touch—sensations dominated his memories: the sleek tailfins on his mother's 1959 Cadillac, the bony backs of the chameleons he kept in a tank in his room, the sleet stinging his face in a winter storm. Sitting in the backseat of the family car on the way to visit relatives in New York City, Sarkin liked picking out the Chrysler Building in the distance. He was entranced, too, by the Brooklyn Bridge, with its shiny, soaring arches and intricate, criss-crossing cables.

Sarkin was also a reflective child, and he still enjoyed thinking about the big questions—the "who" and "what" and "why" of existence—a habit he acquired early in life. He remembers being mesmerized as a child one late afternoon by a slice of sunlight leaking beneath a window shade. The sea of sparkling dust motes revealed by the light astonished him. Obviously, the world was filled with hidden gems, secrets other people rarely noticed.

The middle of three children—Richard was three years older, Jane six years younger—Jon was close to both, but he always went his own way. In 1961, at the age of eight, he wrote to John F. Kennedy in the White House, asking how he, too, could become president of the United States. Ralph A. Dugan, a special assistant to Kennedy, wrote him back: "Study hard and obey the rules. The president is pleased that you ask his advice." But the rules were sometimes unclear to Sarkin or at least pliable. Once, as a teenager, he put a coffee mug down the family's brand-new garbage disposal—it was one of the first in the neighborhood—just to see how well it worked. The machine gobbled the mug up and promptly broke down.

He wasn't an athlete like Richard, who starred in track, soccer, and basketball in high school, but he was just as brilliant. He wasn't an overachiever like Jane, but he read all the time, inhaling Kafka and Kerouac. He was both introverted and unruly, once smuggling a concoction out of high school chemistry class and blowing it up in the middle of the family's driveway. Yet he also volunteered to draw posters and banners for school dances and contributed a cartoon to the high school yearbook. He liked being an adventurer and pushing boundaries, but never too much or too far, because he also liked being popular and fitting in. Mostly, he was torn between being a rebel and pleasing his parents, with the latter usually winning out.

Stanley Sarkin was a hardworking, well-to-do dentist, the son of Eastern European immigrants. He grew up in a New York City tenement during the Depression, highly aware of the great Jewish divide between the rich Guggenheims and poor families like his. After working his way through New York University and then dental school, he joined the navy, married a nineteen-year-old New Jersey girl, and opened his own dental practice in the suburbs. He bought a new Cadillac every few years, joined a golf club, and sent his children to private school. In the summer, the kids went to camp in New York, and for vacation the family visited Beach Haven on the Jersey shore.

Stanley and Elaine were attentive, loving parents, although Sarkin's father could sometimes be distant and judgmental. He played golf with his youngest son exactly once. Jon was fourteen at the time and could barely get the ball in the air, and his father became so frustrated and embarrassed he accidentally drove over his son's foot with the golf cart. They quit after nine holes.

"There are exactly three things that matter in the world," Stanley said to his two teenage sons one day as they sat at the kitchen table talking about the meaning of life. "Money, money, and money." To a boy like Jon, growing up in the heyday of the American countercultural movement, the message was as alien as the tenements of New York. The 1960s and early '70s were dominated by Vietnam, civil rights protests, and political assassinations—John and Robert Kennedy, Medgar Evers, Malcolm X, Martin Luther King Jr. A deep cynicism, even contrarianism, settled into the country, and Sarkin shared the sentiment. Beneath the veneer of middle-class life, he felt there had to be something more, and for him the way to find it was through music: blues, rock 'n' roll, Motown, pop, folk.

In 1962, when he was nine and away at camp for the first time, Sarkin heard the radio blaring in the counselor's cabin. The songs were intoxicating: "Rag Doll," by the Four Seasons; "Surf City," by Jan and Dean; Dion's "The Wanderer." When he finally had a radio of his own, he would spend hours listening to the Beatles, Bob Dylan, the Byrds, Jefferson Airplane, and the Ronettes. His grandfather on his mother's side was in the jukebox business and Sarkin would regularly make lists of the 45s he wanted, which his grandfather would gladly deliver. For variety, he would also play his parents' LPs, mostly show tunes from *South Pacific, The Sound of Music, West Side Story*, and *Oliver.*

Sarkin took guitar lessons briefly in the seventh grade, but he didn't study the instrument in earnest until high school, when a friend taught him the three chords to Dylan's "You Ain't Goin' Nowhere." After that, he practiced all the time, and eventually bought a Stratocaster electric guitar, and then a Gibson lap guitar. In high school, he often took the bus or train into New York City to the famous rock palace, the Fillmore East, to see Procol Harum, Jethro Tull, or Jimi Hendrix, entranced by the Joshua Light Show, whose "liquid light" was projected onto a mammoth forty-foot screen just behind the stage. Sometimes the lights were overlaid with photos or bits of film, and the pictures and music seemed a natural combination to Sarkin. Art spoke to him the same way music did. A friend introduced him to the blues through the album *Fathers and Sons*, by Muddy Waters and the Paul Butterfield Blues Band, and he fell in love not only with the music, but with the album's cover art. It reproduced Michelangelo's mural *The Creation of Adam*, on the ceiling of the Sistine Chapel, but with a black bluesman as God, and as Adam, a white hipster in sunglasses.

From his earliest years, Sarkin liked to draw, mostly pictures of army soldiers or flying saucers, and he enjoyed making movie monsters like Godzilla and King Kong from Erector Sets, or simply rearranging the parts of model cars, gluing the tires to the roof just because he thought it would look cool. Later he was exposed to fine art on family trips into Manhattan to visit the art museums—the Metropolitan, the Whitney, the Museum of Modern Art—and was drawn as much to the Renaissance as to modern art. Rembrandt, Rauschenberg, Botticelli, and the radical conceptual artist Joseph Beuys all appealed to him.

In the fall of 1969, when he was sixteen, Sarkin took the train down to Johns Hopkins to visit his brother Richard in college. At the time Richard was living in an apartment in Baltimore and in his bathroom were Zap Comix books by the artist R. Crumb with the words "Fair Warning: For Adult Intellectuals Only!" on the covers. The stories were ribald and raunchy and deeply countercultural, with characters like Angelfood MacSpade, Whiteman, and Shuman the Human. The crosshatched pen-and-ink drawings included frontal nudity, drugs, and sex and made a deep and lasting impression on Sarkin, not only for their art, but for their subversiveness, which confirmed something his ironic soul had suspected for some time: that there were different levels of reality; that the world was more expansive, more nuanced and certainly less tidy than the one he was familiar with at home and in school. He became interested in cartooning and steeped himself in the underground art of Crumb, "Spain" Rodriguez, S. Clay Wilson, Rick Griffin, and Victor Moscoso. He read gonzo writer Hunter S. Thompson's *Hell's Angels* and *Fear and Loathing in Las Vegas*, with drawings by Ralph Steadman, the illustrator and cartoonist famous for his political and social caricatures. When he was bored in school and doodled in his notebook, he sometimes imitated Crumb's crosshatched style and Steadman's sinister cartoons.

The summer after graduating from high school Sarkin attended a figure drawing class at the famous Art Students League. One day, using charcoal, he drew a very unlifelike figure and the teacher scolded him, saying, "There will be plenty of time to do abstractions like that." Sarkin was unfazed. Whatever he was doing at the time, he wanted more of it—more music, more art, more books. He felt drawn to the creative side of things, always knowing those pursuits could never amount to a serious occupation. His father once told him, "You can do anything you want, as long as it's what I want you to do." For Sarkin, caught between studied reliability and a natural recalcitrance, the message only made him more confused.

As a freshman at the University of Pennsylvania, he remained unsure about his career path, torn again between his artistic impulses and going into medicine, like his father. At first he considered becoming an architect, but then switched to pre-med. He briefly thought about dental school, but finally settled on a degree in biology. When he moved home after graduating in 1975, he was still unsure what he was going to do with his life, but he became intrigued when a relative mentioned environmental studies. Within a few weeks, Sarkin applied for, and was accepted into, the master's program in environmental science at Rutgers.

Shortly thereafter, he moved into a house with four other men, all studying to be chiropractors. They turned him on to vegetarianism and meditation, and before long he was growing a beard and gardening. He read Carlos Castaneda's books, especially the popular *Journey to Ixtlan*. Part travelogue, part mystical philosophy, the book is an assault on rationalism and was a totem of the "let-it-all-hang-out" '60s and '70s.

Sarkin was comfortable among his housemates in expressing a slightly whacky playfulness. When he harvested the basil from their vegetable garden, he gathered up the dried detritus, which looked distinctly like marijuana, and stuffed it into envelopes, scrawling "Season's Greetings!" across the back and mailing them to a number of very surprised friends. In the spring of 1976, he decided to start his own "movement," Beards for the Bicentennial, complete with T-shirts, buttons, and a song he composed. But as much as he enjoyed being unconventional, Sarkin attended all his classes, studied hard, and maintained good grades.

There were no serious romantic relationships until the fall of 1976, when Janet, a former girlfriend of one of his housemates, called him up and asked him out. He liked that she was so self-assured that she called *him*, and they quickly became a couple. Janet was a pretty blonde, four years younger than Sarkin, and an undergraduate at Rutgers. Smart, funny, and perceptive, she was also moody, and right from the start their relationship was combustible. She demanded his attention and would become jealous if he even talked with another woman. He wanted a girlfriend, not a wife, so the relationship caused almost constant tension.

Over the course of his two-year program, Sarkin began to realize that environmental science was not interesting enough to become his life's work. Increasingly he was attracted to the subject his housemates were studying, chiropractic medicine. The discipline has won grudging acceptance by the medical profession over the past twenty years, but its most ardent defenders are patients. For Sarkin, its appeal was twofold. He could be a health professional, but without being a traditional doctor, since chiropractic, which involves the manipulation and readjustment of bones, was not a mainstream science. After he finished his master's degree at Rutgers, Sarkin enrolled in the Palmer College of Chiropractic in Davenport, Iowa, and dove into his studies.

Janet eventually followed him and worked part-time as a picture framer, but their relationship remained rocky. She often picked fights and was jealous of the time he spent with his studies and with his friends. Still, he was determined to make the relationship work. Before graduating from Palmer, Sarkin landed a job working with an established chiropractor in Gloucester, Massachusetts, and shortly after he and Janet returned east in September 1980, they got married and moved to New England.

Sarkin loved being a chiropractor, having someone come into his office hunched over in pain and leave upright and smiling. As a specialty, he chose temporomandibular joint disorder, or TMJ, which involves inflammation of the joint that connects the jawbone to the skull. Somewhere between dentistry and neurology, TMJ specialists were hard to find, especially for patients with intractable migraines. Sarkin found great satisfaction in the work and was particularly gratified when he cured one woman who had suffered debilitating headaches for a decade. Home life with Janet, however, was not going well. She found part-time work as a picture framer again, but she seemed resentful of the time Jon spent at his job and became despondent. Their fighting quickly raged out of control and just a year and a half into the marriage, in April 1982, Janet moved out and filed for divorce. Sarkin was relieved more than heartbroken; he hadn't realized how much she drained him emotionally. Not long afterward, he met Kim Richardson.

Kim was twenty-five, working as a teacher's aide and librarian at a grammar school in Hamilton, Massachusetts, and every Tuesday and Friday nights she supplemented her income by waitressing at the Blackburn Tavern in nearby Gloucester. After a falling out with his boss, Sarkin started his own practice in Hamilton, and on his way home sometimes stopped at the Blackburn for a bite to eat. When he noticed a pretty blonde waitress there one day, he began making it a point to come in for dinner whenever he thought she was on duty.

"There's that guy again," Kim told a fellow waitress in the kitchen one day. "I know he's going to ask me out."

The other woman went out to the dining area to take a quick peek at the lanky, slightly preppy man sitting at the bar. When she returned she placed her hand on Kim's forehead, as if testing for a fever, and said, "Are you kidding?" Sarkin just didn't strike her as right for Kim, but Kim was intrigued by him; he was serious and polite, though he hadn't said more than ten words to her in the dozen or so times she'd waited on him. Finally, after several weeks, he made his move.

"Do you want to come by and see my office sometime?" Sarkin asked.

Kim was startled by the question.

"Why would I want to do that?" she said, laughing, then added, more gently. "If you like, why don't we get a drink or something to eat?"

A few days later, he picked her up in his shiny new Saab and they had dinner at a nearby jazz club. Kim was impressed: He didn't drink or smoke and he actually owned his own car. He wasn't exactly outgoing and rarely spoke about himself, but she found his quiet reserve, even his old-school etiquette—the way he pulled a chair out for her when they sat down at a restaurant—very appealing. He could also be slightly unpredictable and often left her notes on her car: "See you at 6," or "Meet you at the tavern."

Just a few years earlier Kim had returned to Cape Ann after graduating from the University of Massachusetts Amherst. She was restless being single. She didn't want to hang out with friends and talk about men, or worry whether some boyfriend was attentive enough. She just wanted to get on with it—to find someone with whom she was compatible, marry him, and have children.

At the time, Kim was considering moving to South America, where she had spent a semester in college teaching at an American school. Teaching had struck her as the perfect job, because she could surround herself with kids before she was married, and if she had to keep working when she did get married and had children of her own, she would be able to get home from school when they did.

Kim had already filled out applications for several jobs in Venezuela, Colombia, and Ecuador, but now she put the plan on hold. Her relationship with Jon was quickly becoming serious, and in June 1986, they were married.

By that time, Sarkin's practice was thriving and when they returned from their honeymoon, he decided to build a house, helping to design it and then contracting all the work. Tucked into the side of one of Gloucester's many hills, the wood-frame home sat on eight sloping acres, with a vernal pond at the bottom of the backyard and a large greenhouse, which Jon filled with cactuses, his favorite plant.

When he was a child, Sarkin loved reading Dr. Seuss books and one of his favorites was *If I Ran the Circus*, the story of a man named Mr. Sneelock who "roller-skis" through a slalom course of cactuslike "stickle-bush" trees. When a young Sarkin asked his mother about the strange plants, she told him they were prickly and dangerous to touch, and that was all a contrarian child needed to know.

Kim wanted to start a family right away, and Jon was all for it, so when she became pregnant nine months later they were both thrilled. What Kim wanted most in life was to become a mother. From a very young age, she loved children and knew that she wanted to have lots of babies when she grew up, even though that was not exactly the thing to admit during the feminist revolution of the 1970s and '80s. When she was three years old, she found a Chatty Cathy doll under the Christmas tree that she treated just like it was her own child, keeping it with her wherever she went. Today the doll still sits on a shelf in her closet.

Her life with Jon fulfilled her dreams. Curtis was born in a birthing center with a midwife, and Jon was at Kim's side the whole time.

When they brought him home, he slept in their bed, not a bassinet, and Kim and Jon liked to joke that Curtis's first real piece of baby furniture was a king-size bed. They'd plop him down in the middle of it and call him "King Curtis."

When Kim was awake, the baby rarely left her hip. She had become a fervent believer in attachment-style parenting—keeping your baby as close to you as possible—after reading *The Continuum Concept* by Jean Liedloff right before Curtis's birth. Liedloff had traveled to the Amazon jungle, in the Caura River basin of Venezuela, and observed an indigenous tribe of people there, the Tequana, who believed in "natural" child rearing. In *The Continuum Concept*, she detailed how mothers in the tribe held their pre-crawling children as much as possible and slept with them in the same bed. Kim had decided that was the way she would raise her children. She set up a changing station right next to the bed, and even brought the baby into the shower with her.

Parenting was also deeply satisfying for Jon. Years later, when Curtis was a teenager, playing the part of the stage manager in a high school production of *Our Town*, Sarkin was particularly moved by one of his son's lines, that "every child born into the world is nature's attempt to make a perfect human being." For Sarkin, Curtis's birth was a challenge to become an exemplary parent. It was perfectly clear in his mind that the most important thing he could do was provide for his family, just as his father had done, and he immediately started putting money into a college fund for Curtis. The stability he found with his wife and son was profoundly pleasing to him. They were a family and they were building their lives together.

One night, when Curtis was a few months old sleeping beside his mother, Sarkin bent down and kissed his sleeping son on the forehead, then ran around to the other side of the bed and hopped in.

"Isn't it amazing how happy we are," he said. "It's almost scary."

In August of 1989 they had finally come through an ordeal they could never have imagined a year earlier they would face. They'd met it head on, they had won the fight, and they could finally get their lives back on track.

Chapter 6

CODE BLUE

The day after deep-brain surgery, Sarkin dipped in and out of a drowsy half-sleep. Still woozy from the operation, he was well enough to sit, propped up by a levee of pillows. Bandages were wrapped around his head covering the wound in his skull and the crescent-shaped scar behind his left ear. His mother, Elaine, was sitting with him when, early in the afternoon, Kim walked back into the room with the baby in her arms. She had taken Curtis to the hotel for a couple of hours. Sarkin turned toward his wife and child and opened his eyes.

Kim was taken aback. "Something's different," she said quietly to Elaine.

She noticed it right away: a distant, glassy look in her husband's usually warm green eyes.

"Jon?"

Suddenly Sarkin's face broke into a strange half-smile. He looked at his mother and patted the covers of his bed.

"Come here, Ida," he said, looking straight at Elaine.

Kim took a sharp breath. Ida was their black Labrador retriever. She rushed from the hospital room into the hallway.

"We need help!" she called out. "We need a doctor!"

A young resident standing at the nurse's station dropped his paperwork and walked briskly into the room.

"How are you doing, Mr. Sarkin?" he asked in a voice just a notch above normal.

"Come . . . here . . . Ida," Sarkin said again to Elaine, this time garbling the words.

The resident shone his penlight into one eye, then the other, hoping to see them dilate, a sign that the basic functioning of the brain is intact. The test is a check for whether there is bleeding in the brain. The cranial nerve that controls constriction of the pupils is located high on the brain stem. If there is pressure on that nerve from a buildup of blood leaking, say, from a ruptured vessel, then the pupils fail to react. And if they fail to react—that is, if the pupils are "blown"—there is only one conclusion: something catastrophic has happened inside the patient's head.

Sarkin's pupils barely moved. The resident quickly peeled the bandages from his head and as the last bit of gauze was pulled away, he visibly paled.

"There's a lot of blood," he said to no one in particular.

Then, without looking at either of them, he said to Kim and Elaine, "Please step out."

Neither moved. They were still trying to understand what was happening.

"Please step out *now*!"

Sarkin had suffered a massive stroke. Somewhere at the bottom of his brain a blood vessel had burst, soaking the cerebellum in blood and causing it to swell. His brain was pressing so hard against his skull that blood had leaked through the surgical burr hole in the bone and breached the dam at the site of the wound in his scalp. Now the cerebellar tissue was pushing down on the spinal cord, cutting off the flow of oxygen-rich blood into the brain stem. Any second Sarkin might go into cardiac arrest and lose consciousness.

He stopped breathing. Unless something was done in the next few minutes to relieve the pressure, Sarkin would almost surely die.

The resident punched an emergency code into the phone, and a few seconds later there was a hospital-wide announcement.

"Code blue! Code blue! Code blue!"

The call could be heard all the way down the hall in the waiting area, where Kim and Elaine had retreated with the baby. Seconds later, a crash cart and a half dozen emergency personnel barreled by them. They worked on Jon for a minute in his room and then rushed him toward the elevator. With all the medical people surrounding the gurney, it was difficult for Kim and Elaine to see Jon, but they could see enough to tell that he wasn't moving, and that a doctor was giving him chest compressions.

Elaine began to cry quietly and Kim paced the floor, clutching Curtis tight to her chest.

"Please God, please God, please God," she whispered as she walked in circles.

After a few minutes, a nurse came in and explained what was happening. There was swelling in Jon's brain, which meant there was a blood clot, and he had been taken into surgery to relieve the pressure against his skull and find the source of the bleeding. Kim and Elaine asked why this was happening, what had gone wrong, but the nurse wasn't able to give any answers.

The two women each retreated into her own thoughts. For Elaine, who'd lost a husband so suddenly and unexpectedly nearly twenty years earlier, it didn't seem possible she could now lose a son. For Kim, the idea of Jon dying was too surreal to even contemplate.

A few minutes later, a priest entered the hospital room where Kim and Elaine were now waiting. Kim physically recoiled. A priest comes only to comfort the grieving, she thought. He was an omen that her husband was going to die.

"No, no, no," she said to him. "I don't want to talk to you."

When Dr. Jannetta joined the emergency surgical team in the operating room, he cut into the wound he had stitched up just twenty-four hours earlier and used the air drill to widen the opening in Sarkin's skull. It took him just five minutes to reach the cerebellum. Finding the clot amid all the blood, and removing it, would take much longer. Somewhere in that jungle of blood vessels threading in and around the cerebellum, one of them had ballooned and sprung a leak. Locating it would be like trying to find a small hole in a boat already under water. The team worked feverishly, and as Jannetta delicately probed the deep brain tissue, one of the monitors in the room suddenly let out a high steady whine. Sarkin had flat-lined. For a second time he was in cardiac arrest. One of the doctors reached for the defibrillator paddles.

"Clear!"

A few seconds later 5,000 volts of electricity penetrated Sarkin's heart muscle, stopping its irregular rhythm and then restarting it. He was revived, but not before tiny pockets of oxygen-starved brain tissue had begun to die.

Jannetta was shaken. He had never lost a patient doing this procedure. In fact, his closest confrontation with death had been his own. He had been a freshman at the University of Pennsylvania when he noticed a bulge in his neck, just under his chin. For months he paid no attention to it, but eventually he could no longer button his collar. Finally, he decided to have it checked, and surgery was recommended. He underwent a lengthy operation. When he woke up, the bulge was gone, but he could barely talk. The doctors told him they had removed part of his thyroid, a butterfly-shaped gland just below the larynx, and that the surgery had irritated his vocal cords, but they said he would make a full recovery.

Five years later, in his second year in medical school, Jannetta became curious about the surgery and obtained the tissue samples from his operation. He asked a couple of his teachers to examine the specimens and what they found shocked them. The tissues revealed thyroid cancer, and over the years, it had probably spread. Thyroid cancer is extremely rare in young people, and even when a nodule is discovered, 99 percent of the time it is benign. Unchecked, however, the cancer can spread to other glands, surrounding tissue, and blood vessels. Jannetta's teachers could not understand how the cancer cells went undetected the first time around, and five years without any treatment was more than likely a death sentence. They told him he should decide what he wanted out of life, because he'd likely be dead within a few years. After more surgery, and much medication, Jannetta's thirtieth birthday came and went. Then his fortieth. Then his fiftieth. He doesn't remember when exactly, but at some point, he stopped looking over his shoulder and settled into his life and career.

Now, working feverishly over Sarkin, Jannetta was unsure for the first time in thirty years whether a patient of his would make it. Usually if something went wrong after a microvascular decompression, it happened during the procedure, not a day later. A delayed problem like this, with the patient suddenly becoming so sick? That just didn't occur. But here he was, probing the most sensitive structures of his patient's brain for a second time.

"It's where we live," he often said about the brain stem. "It's an unforgiving area."

Jannetta used two probes, one in each hand, to gently part the layers of brain tissue in search of the tiny, jellylike clot. His mind raced through the possible causes of the stroke. Could he have nicked a blood vessel or damaged the cerebellum when he moved it out of the way? Jammed up against the skull on one side with the brain stem below, the left side of the cerebellum was already dying. Jannetta would have to carve away some of the dead tissue to relieve the pressure and he would have to do it fast. If it continued to swell, pressing harder and harder against the brain stem, it would soon kill Sarkin.

Peering intently through the lens of the standing microscope, Jannetta quickly but carefully removed layer after layer of Sarkin's cerebellum, worrying about the damage he was doing with each bit of tissue he tore away. The cerebellum controls motor coordination, posture, and balance and helps organize and make sense of visual images.

But there was more. Jannetta knew damage to the cerebellum could also affect cognition, emotion, and behavior. Surgery in this part of the brain has rendered some patients mute; others have developed depression, an autisticlike intolerance of others, or an inability to empathize. What would be left of Sarkin after all that brain matter was removed?

Finally, Jannetta found the tiny clot and plucked it out. After nearly three hours of surgery, he left the OR and collapsed in a chair at the back of the surgeon's locker room. He sat by himself for a few moments, then picked up the phone. He dialed his mother at home in Philadelphia, telling her about what had happened, that his patient was very sick and might well die.

"Pray for him, Mother," he said.

In the intensive care unit, a machine was breathing for Sarkin and he was in a light coma. Hours had gone by since he had been rushed away when a nurse finally came to tell Kim and Elaine that Jon was out of surgery. His brother, Richard, and sister, Jane, were there now, too.

Kim and Elaine's last image of Jon had been of his lifeless body on a gurney hurtling down the hall. They'd spent most of the time since then drinking coffee and watching Curtis play on the floor of the waiting room or walking him up and down the hallway. When they were told Jon was in recovery they almost didn't believe it.

"How is he?" they asked at the same time. "Is he okay?"

The nurse told them the surgeon would come to talk to them soon and that they should go back into Jon's hospital room to wait for him. Finally, Jannetta joined them, still in his scrubs. Even twenty years later, Kim remembers the moment as if time had paused and etched the images in glass. Jannetta slowly walked in, leaned his back against the wall, and then slid down to the floor, briefly putting his head in his hands. What a curious coincidence, Kim thought; her husband had done exactly the same thing in the hallway of their home that day so many months ago.

"I don't know what happened," Jannetta finally said, looking up at Kim and Elaine.

"Will he recover?" they asked.

Jannetta had no good answer.

"He might live or he might die," he told them. "Or he might be something in between."

Chapter 7

THE MECHANICAL MIND

E ven asleep, the human body is never at rest. The heart flutters, the blood whispers, the lungs swell and subside, and beneath the hard plates of the skull rivers of electricity ignite our dreams. For centuries, natural philosophers and then scientists struggled to understand the source of our life force, our consciousness, the animating spirit of our soul—in short, what makes us who we are. Eventually, the brain came to be understood as the seat of the self or the soul, though the manner in which it gives rise to the "I" we all perceive ourselves to be is still an unfolding mystery.

The pioneering English physiologist Charles Scott Sherrington lyrically expressed the wonder of the brain's workings when he wrote in 1942 that, awakened from sleep, the brain "becomes an enchanted loom where millions of flashing shuttles weave a dissolving pattern, always a meaningful pattern though never an abiding one . . ." What we see, hear, feel, and experience of the world, all of it Sherrington believed, was mediated by the three-pound mass of gelatinous tissue in our heads. How the brain accomplishes this is only now being pieced together by neuroscientists. The magical part of consciousness—how, and even why, a sense of self arises from these complex perceptions—is still an enigma.

No one, least of all the sleeping Sarkin, knew whether he would be the same man when, and if, he awoke. How much, and what parts of his brain, would remain? And would they be the parts he needed in order to understand who and what he now was?

The brain has not always inspired the respect Sherrington showed it. In 1652, for example, British philosopher Henry More viewed the brain with outright disdain: "This lax pith or marrow in man's head shows no more capacity for thought than a cake of suet or a bowl of curds." The ancient Egyptians believed the brain was purposeless, except as a kind of stuffing or packing material for the skull. Deemed inessential, the organ was removed from deceased Egyptians before burial, usually by extracting it through the nose. Other organs, such as the liver, kidneys, and bowels, which were thought to house a person's desires, emotions, and personality traits, were often kept in jars. Only the heart remained inside the body when it was entombed. To the ancient Egyptians as well as the Hebrews, a person's life force was generated by the heart, which throbbed beneath the chest, rising and falling with each breath. Aristotle, too, located man's vital spirit in the heart.

For most of civilization, the insubstantial essences of thought, personality, and emotion were believed to emanate from the physical corpus. Our modern-day vocabulary even betrays these ancient roots. When we tell the truth we "speak from the heart," when we're angry our "blood boils." In times of danger, we "gird our loins" to muster our courage because it "takes guts" to prevail over our enemies. Our hearts "bleed" in sympathy with the problems of others and "ache" for our own loss when we mourn the death of another.

Not until the second century was the status of the brain upgraded by the Roman physician Galen. Air, or pneuma, he suggested, was the essence of the life force, and it took three forms, each with a different function. The vital spirit was located in the heart and controlled blood flow and body temperature; the natural spirit was situated in the liver and was responsible for nutrition and metabolism; and in the brain resided the animal spirit, which accounted for sensory perception and movement. The structure responsible for converting the vital spirit into the animal spirit was an area of dense blood vessels, called the *rete mirabile*, or miraculous net, located at the base of the brain. In many mammals, this vascular network is found in the neck or even the limbs and helps control body temperature. According to Galen, the actual tissue of the brain had no real purpose, but in this net of blood vessels was stored a person's animating spirit, which traveled out into the body through the brain's hollow nerves.

Astonishingly, Galen's ventricular theory of brain function persisted for nearly 1,500 years. Motion, memory, reason, sense impressions—all were attributed through the centuries to the animating spirits transported throughout the body by the ventricles, which originate in the brain. Galen's proof? He had observed the *rete mirabile* in every dissection he'd performed. There was only one problem: because human dissections were prohibited by the Roman church, Galen's observations were based almost entirely on dissections of cows and sheep.

Although the first recorded human cadaver dissection dates to the sixth century B.C., religious bans and widely accepted superstitions made them rare until Flemish pathologist Andreas Vesalius revived the practice. Born in 1514, Vesalius showed an early interest in anatomy, dissecting rats, stray dogs, and cats. At the age of nineteen, he entered medical school at the University of Paris and quickly became bored with the habit of professors reading to students from the ancient texts of Galen. So did Galileo, a medical student at the University of Pisa in 1581, who promptly switched his course of study to mathematics and mechanics. Vesalius, however, found a different solution for his boredom. He rummaged through the bones of the dead in the charnel houses and sneaked into Paris's Cemetery of the Innocents to dig up the bodies of executed criminals, all in order to study the workings of the human body.

Leonardo da Vinci's riveting anatomical drawings, based on his own cadaver dissections, helped to ease the public opprobrium, so that by the time Vesalius graduated and accepted a teaching position at Italy's University of Padua, he was able to acquire a steady stream of cadavers from criminals executed by the courts. In 1543, the same year Copernicus declared the earth orbited the sun, Vesalius published *De Humani Corporis Fabrica* (On the Fabric of the Human Body), a 663-page anatomy text, which also included this important conclusion: Galen's "miraculous nets" might be part of the brains of other animals, but in humans they didn't exist. Vesalius wrote that he was "completely astonished" at his own "stupidity and too great trust in the writings of Galen."

In 1564, the year of Vesalius's death, the Reformation deposed Catholics from rule in England, and the Royal College of Physicians of London lifted the ban against human dissection. At the time, most philosophers and scientists still believed that everything from diseases and disorders to personality traits and emotions could be explained by the ebb and flow of the body's four fluids: blood, phlegm, black and yellow bile. The brain was still a bit player in the drama of the self until the French philosopher and mathematician René Descartes helped to elevate its status by making a distinction between immaterial thought and physical substance. His theory was grounded in his famous dictum, *Cogito ergo sum* or, "I think, therefore I am." Descartes argued that if he could think about himself—even if his thoughts were to doubt his own existence—then there was a self doing the thinking or doubting. Personal existence, in other words, could be established through logic. The body, on the other hand, was a physical substance in time and space, a complex mechanism that walks, talks, sees, hears, eats, and sleeps. Having proven his own existence by argument alone, Descartes then asked himself the mother of all follow-up questions:

"What is this 'I' that I know?"

Of all the animals on the face of the earth, only one—*Homo sapiens*—is capable of asking the question. For Descartes, the answer was the soul, which encompassed the whole of human passions, perceptions, and sensations. This soul, he argued, was located in the tiny pineal gland, an endocrine organ whose hormonal functions were identified only in the twentieth century. The soul, he asserted, occupied a kind of middle ground between body and mind, and the brain's pineal gland acted as a kind of way station between the two.

Unfortunately, before he died at the age of fifty-three tutoring Sweden's Queen Christina behind her drafty castle walls, Descartes never got around to explaining how body and mind interacted in the pineal gland. In a letter dated June 28, 1643, a little less than seven years before his death, he admitted finding such an explanation was probably a lost cause, since "what belongs to the union of the soul and the body is known only obscurely by the intellect alone or even by the intellect aided by the imagination . . ."

Ten months after Descartes's death, English physician and anatomist Thomas Willis witnessed an astonishing medical event that ultimately confirmed for him the union of soul and body. On Saturday, December 14, 1650, in the prison yard just outside Oxford Castle, England, a twenty-two-year-old servant girl named Anne Green was to be hanged at dawn. Green had been found guilty of fornication and the murder of her infant. At the time, she was employed in the household of Sir Thomas Reade in the village of Dunstew, some sixteen miles north of Oxford on the edge of the Cotswolds, and it was Reade's teenage grandson, Jeffrey, who probably got the servant girl pregnant. After suffering abdominal pains, Green, who may not have even known she was expecting, spontaneously aborted the fetus in an outhouse. Not realizing she had actually miscarried, Green retired to her room, sick to her stomach, and lay down on her bed. Later, the bloody sheets alerted others that she might have prematurely delivered and a search uncovered the infant's remains in the outhouse. The fetus was some nine inches long, and was probably miscarried in about the twentieth week. Three days later, Green was arraigned at the castle yard in Oxford by Umpton Croke, who was both sergeant of the jail and judge. Without delay, Croke pronounced Green guilty and sentenced her to hang.

Willis, along with his colleague William Petty and several other Oxford scholars, planned to dissect the young woman's body. Petty held the chair of Tomlins Reader in Anatomy at Oxford, and according to a royal order issued fourteen years earlier by Charles I, the Tomlins Reader was entitled to the body of anyone executed within twenty-one miles of Oxford. The bodies were needed not just for the general study of human anatomy, but for medical students who were required to participate in two public dissections before receiving their degrees.

Three weeks after Croke sentenced Green to death, the stout young woman was escorted from her cell into the yard. A hymn was sung by those present and two ladders placed against a simple wooden post and beam. Green ascended one ladder, her executioner the other. Before the sentence was carried out, the young woman spoke for some time, professing her innocence and decrying the abuse she'd received at the hands of both Thomas and Jeffrey Reade. When she finished, the executioner, draped in black, pushed her off the ladder. For those hanged in this manner, death by asphyxiation was slow and agonizing. To hasten it, those in the crowd would often pull on the criminal's legs, and they did so for Green, as she had requested before being hanged. After half an hour her lifeless body was cut down and placed in a wooden coffin, which was then taken less than a mile away to 107 High Street, where William Petty rented rooms.

For the dissection, Petty and Willis were joined by several other members of the informally organized Oxford Experimental Club, which years later would become the Royal Society of England. The Philosophical College, as they sometimes referred to themselves, also included student Richard Lower, who eventually performed the first blood transfusion; chemist Robert Boyle, whose experiments with gases led to the definition of a chemical element; and Christopher Wren, the mathematician and architect who also pioneered the intravenous injection of drugs. Boyle also provided an important tool to the science of dissection when he discovered pure alcohol could preserve human tissues and organs as well as animal specimens. In

1650, however, speed was imperative in human dissections, and in the study of brain tissue in particular, since neural cells decompose within three to seven minutes postmortem.

When Green's coffin was opened sometime around 9:00 a.m., the noose still hung around her neck. As the group gathered around the body, Green's chest seemed to rise, ever so slightly, as if she might still be breathing. Startled, one of the men who had carried the coffin into the room began to stomp on the woman's chest to finally put her out of her misery. Moments later Petty and Willis arrived and stopped the man from further abusing Green's body. Then, hearing what they thought was a rattle coming from her chest, they rushed to revive her, first propping her up and then prying open her mouth and pouring liquid down her throat. Green's face appeared swollen and flushed. Willis and Petty placed tourniquets around her limbs and massaged her fingers, hands, arms, and feet. When they tickled her neck with a feather, she briefly opened her eyes. The efforts at resuscitation continued with bloodletting, an enema, and finally putting her to bed beside a healthy woman to keep her warm. At around six o'clock that night Green was well enough to speak a few words, and by Sunday she could answer questions.

When the physicians asked what she could remember of Saturday's events, she talked about giving away some of her clothes to her mother in the hours before the execution, but she had no recollection of being escorted to the gallows, the hymn, or anything she had said before the noose was placed around her neck and she was hanged. Instead, what the gentlemen scientists observed, was that "she came to her self as if she had awaked out of a Sleep, not recovering the use of her Speech by slow degrees, but in a manner all together, beginning to speak just where she left off on the Gallows . . . She remembered nothing at all that had been done unto her," wrote Richard Watkins, who was present in the room, and she seemed merely "to go on where she had so long time left off; like to a clock whose weights had been taken off awhile, and afterwards hung on again."

By December 19, Anne Green was out of bed and eating roast chicken. A few days later, with the help of testimony by members of the Oxford Experimental Club, she was pardoned by the court and returned home to her family, taking her coffin with her as a memento of her "resurrection."

For Willis the physician, the event, which became widely publicized, set him up for life. Born in 1621, he had entered the University of Oxford at age sixteen intending to study theology, but medicine quickly overtook his curiosity and when he received his medical degree in 1646, he set up a practice in Oxford. Short in stature, with dark red hair and a tendency to stammer when he spoke, the young Englishman had trouble attracting patients to his new practice. He was forced to share the use of a horse, and sometimes advertised his services in neighboring markets. Green's "resurrection," however, turned him into a local celebrity. Poor and rich patients alike poured into his practice. He often treated the former for free, and for the latter he was perhaps the first doctor to use a sliding scale to determine payment. According to tax records from that time, Willis eventually became Oxford's wealthiest citizen, earning the equivalent today of three-quarters of a million dollars a year.

Like Vesalius, Willis performed dissections, not only on animals, including horses, dogs, monkeys, lobsters, oysters, and earthworms, but on scores of humans. His comparative dissections enabled him to distinguish differences in the anatomies of mammals, reptiles, fish and arthropods, and also to identify what was uniquely human. Willis's research methods were impeccable. For one thing, he did what no doctor or scientist before him had ever done: he not only treated his patients, but observed and studied them over long periods of time and then dissected their brains after death. In this way, he was able to correlate symptoms he observed in his patients while they were alive with the deformities and deficits he observed in their brains postmortem. Willis's understanding of brain structure was still primitive. In his rough-hewn anatomy, the medulla oblongata took in the whole of the modern-day brain stem, and the corpus callosum, he believed, included not just the white matter that connected the left and right hemispheres, but much of the higher gyri, or folds, in the neocortex.

Despite his misreading of brain structure, Willis was one of the first to tie specific functions of the nervous system to particular areas of brain tissue—not to the fluid of the ventricles. He hypothesized that all voluntary motion originated in the striatum, located on the lower lateral wall of each hemisphere. The corpus callosum was responsible for fantasy and imagination, and memory and sensation were products of the interaction between the corpus callosum and the cerebral cortex, the higher brain regions associated with cognition. Willis even had a cerebral explanation for nightmares. Because they involve a sense of helplessness and immobility that caused stressed breathing and heart palpitations, he located them in the cerebellum, which we now know not as the source of breath and respiration, but as an important structure in balance and coordination.

To understand how revolutionary his ideas and methods of study were, one must remember that modern medicine had barely begun at the turn of the seventeenth century. Neither bacteria nor the existence of the cell had yet been discovered, an understanding of the circulation of blood would take another twenty-five years, and the first medical use of the microscope was still decades away. Even the invention of the thermometer and the stethoscope would take another hundred years. In the meantime, death from flu, famine, scurvy, and plague was common, and theories of health primitive: bathing was discouraged because the body's natural oils were thought to provide protection from disease, brain size was believed to be affected by the phases of the moon, and phlegm was considered to be a product of eating too much fish. As far as psychiatric disorders (mania, melancholia, and hysteria), the culprit was usually thought to be the liver, the spleen, or the uterus—never the brain.

Willis, however, believed otherwise, and only through dissections, he wrote, could he unlock what he called, "the secret places of Man's Mind." Often he directed these dissections, allowing Richard Lower, one of his best students, who was particularly skilled in surgical technique, to remove the brain through the back of the skull. The organ was then sliced into numerous specimens, which were studied using a magnifying glass. Lower and Christopher Wren, who also sketched each specimen in detail, sometimes injected dye into the brain, which is how Willis discovered the flow of blood through the cerebral arteries and the "circle" of arteries at the base of the skull that now bears his name. He also found out that the cranial nerves were not hollow, as had been thought, and that nerves create muscular movement. Credited with establishing the concept of neurology and even coining the word, Willis was also an early believer that mania, depression, mental retardation, epilepsy, and narcolepsy were all disorders of the nervous system.

The brain was nothing more or less than a machine, and anatomy, not alchemy or mythology, drove his beliefs about the soul as well. The brain, Willis argued, was the seat of sensation and movement as well as intellect and imagination, but all are enabled by the soul. Like Descartes, Willis believed man's soul resided in the brain, but instead of the tiny pineal gland, he pointed to the animal spirits that travel like impulses throughout the brain—a theory that bears striking similarity to the chemistry of neurotransmitters. For Willis, studying the brain was like looking into the "living and breathing Chapel of Deity."

Anne Green's "resurrection" went a long way to confirming for Willis that the mind was a product of the brain. More than a decade later he would put his beliefs in writing in his textbook, *Cerebri Anatome*, basing them on his experiments and his dissections, but also on the curious, and miraculous, case of Anne Green. At dawn on December 14 she had proclaimed her innocence and railed against the Reade family just before her execution. At nine that morning, she was "resurrected" with the help of Willis and Petty and essentially picked up where she left off, as if nothing had happened in the intervening hours. With "death" her memory stopped, and when she was revived, so too was her capacity to remember. Like a spigot, her mind had been turned off, then turned on again. The brain was a material object, a device that, when its parts were all working, could think, speak, and remember; it was neither a bowl of useless curds nor a mysterious conveyor of animal spirits. In this new age of experiment and empiricism, the brain was a mechanism that could be changed by illness or disease, broken by trauma, and shut down by death. But as the scientists who surrounded the breathing body of Anne Green realized, it could be revived and healed as well.

Chapter 8

Two Halves

L ike every other mechanical device, the brain is susceptible to breakage and failure. The idea that brain injuries can cause loss or alteration of the senses, movement, speech, and intellect dates at least as far back as the first century, when Galen treated Roman gladiators and witnessed firsthand the myriad effects of traumatic brain injuries.

The fact that specific parts of the brain, and in particular its two halves or hemispheres, could account for specific abilities was not widely understood, however, until the middle of the nineteenth century. The discovery of hemispheric function, known as lateralization, is usually attributed to the Frenchman Paul Broca who, on April 2, 1863, published a seminal paper describing patients who had suffered damage to their left hemisphere, and in particular their left parietal lobe, leading to a loss of speech, the ability to write, or the understanding of language.

Remarkably, a week earlier, another Frenchman, physician Gustave Dax, had submitted a paper that claimed his father, Marc, had also postulated the idea of brain lateralization—twenty-five years earlier—in a speech before the medical society of Montpellier, in southern France. Marc Dax's "Lesions of the Left Half of the Brain Coincident with the Forgetting of the Signs of Thought," apparently written in 1836, recounted his discovery of brain lateralization. Unfortunately, he died the following year at the age of sixty-seven before he had a chance to publish his research.

Dax's findings were important. They were based on more than a decade of observation and treatment of patients, as well as on the study of the medical literature. Just a country doctor in the village of Sommières, next to Montpellier, Dax credited three cases of aphasia, or loss of speech, between 1800 and 1811, with confirming his belief in the lateralization of brain function. The first case was a former cavalry officer with whom Dax had become acquainted. The captain had been wounded in battle by a blow from a saber, causing him to lose his memory for words. When the man died a short time later, Dax asked the family for more details about the original injury and learned that an autopsy had uncovered damage to the officer's left parietal lobe.

In 1809, Dax visited a patient with a cancerous tumor on the left side of his face and an impaired memory for words. Finally, in 1811, he read about the death five years earlier of the prominent French naturalist Broussonnet, who had suffered an "apoplexy," probably a seizure or stroke. On autopsy, a large "ulcer" was found on the left side of Broussonnet's brain. Dax began to think about the cavalry officer and the man with the facial tumor, along with the evidence from Broussonnet's brain, wondering if he might be correct about hemispheric brain function.

Every year thereafter, Dax seemed to come across another case until he'd compiled more than forty of them. One of the last was a woman who fainted and fell from her chair. A family member rushed to get Dax, but by the time he reached the patient, she had regained consciousness. The woman admitted to him that while she was unaware of what was happening for only the briefest of moments, she couldn't speak for the first couple of minutes after awakening. "These words were a ray of life to me," Dax later wrote. Two days later, he was called to the woman's house again. She had suffered another attack, only this time much worse, and she was now mute. Dax knew immediately what to do: he applied leeches to the woman's left temple, and after a few minutes she was speaking again.

At the time, most physicians regarded bloodsucking leeches as an important medicinal tool because too much blood was considered unhealthy for humans. Bloodletting dates back to Mesopotamia and may have remained an important medical procedure because reducing blood volume does in fact reduce blood pressure, and thus is especially useful for hypertensive patients. In 1833 alone, France imported more than 40 million leeches from Russia for medicinal purposes. An added benefit was revealed in 1884 when British physiologist John Berry Haycraft discovered that leeches also release an enzyme called hirudin, which acts as an anticoagulant.

Today, in fact, leeches are sometimes used to restore blood to reattached limbs, ears, even scalps, and some homeopathic practitioners recommend leeches as prophylactics against stroke much the way aspirin and heparin are used to thin the blood. For Dax's stroke patient, the leeches increased blood flow to her left hemisphere, which is why she was able to speak again. Within half an hour, she was completely recovered. For Dax, the woman's case reinforced his belief in brain lateralization. "I believe it possible to conclude not that all diseases of the left hemisphere necessarily impair verbal memory," he later wrote, "but that, when this form of memory is impaired by disease of the brain, it is necessary to look for the cause of the disorder in the left hemisphere, and to look for it there even if both hemispheres are diseased."

Years later, when Broca read Gustave Dax's report about his father's Montpellier talk, he scoured the medical literature in search of a written copy. He even contacted the librarian of the Montpellier Medical Society who, on Broca's behalf, tracked down twenty physicians who had been present at the meeting of the society in 1836. None of them recalled Dax's presentation. In 1877, Broca finally found Dax's original manuscript and believed it to be genuine, although he also believed Dax never gave the talk. Perhaps Dax had gotten cold feet because he couldn't support his findings solely with autopsy reports, but he had also relied on observation and anecdotal evidence. Broca was more methodical, which is perhaps why he wrote, in 1877, "I do not like dealing with the questions of priority concerning myself. That is the reason why I did not mention the name of Dax in my paper."

More recent research in hemispheric function has confirmed lateralization not only in humans but also in other animals. Monkeys, like *Homo sapiens*, express fear more strongly on the left side of their face—the muscles on the left side of the face and body are controlled by the right hemisphere, and vice versa—so when the right hemisphere reacts to danger, the left side registers the emotion. Likewise, chickens, lizards, toads, and baboons respond more quickly to a predator when it approaches them from the left side. Again, the view from their left eye is controlled by the right hemisphere of the brain, which responds to attack more rapidly and strongly than the analytic left hemisphere.

Even as Dax and Broca were detailing how language and speech arose in the brain, the idea that a person's very nature or personality could be grounded in brain tissue seemed wildly improbable, if not impossible. Fifteen years before Broca published his seminal paper on lateralization, however, a catastrophic accident to a railroad worker deep in the woods of Vermont led at least one country physician to believe otherwise.

In the fading New England afternoon of September 13, 1848, a horribly injured young man sat on the porch of Adam's Inn with his head in his hands. Twenty-five-year-old Phineas Gage was fully conscious and talking, even though a sliver of his brain protruded from a jagged opening in his skull, dangling in the daylight for all to see.

"Doctor, here's business enough for you," Gage said to Edward Williams, one of the town's two physicians, as Williams drew close.

Ten minutes earlier, the doctor had been intercepted on his way home by a railroad hand, a member of Gage's crew, which had been clearing rock just south of the town of Cavendish, Vermont, for the new Rutland & Burlington line.

"He's hurted," the man said about his foreman, "and he's waiting for you at the hotel beyond."

Just twenty-four, Dr. Williams was younger than Gage and only recently out of medical school. He had never seen such a wound before. A patch of bloody brain tissue was hanging out of a large opening in Gage's skull, and it was pulsating.

Gage was a popular and hardworking foreman. He was good-looking, a fair-haired, sturdy 5'6", with a strong jaw, a slight cleft in his chin, and light blue eyes. The young laborer had grown up on a farm in Lebanon, New Hampshire, less than forty miles to the north, and he had been working on the construction gang in and around Cavendish for several weeks, blasting apart granite rocks and leveling the earth in order to build a bed for the rails. Twenty-five miles of tracks already stretched north from Bellows Falls, where the line began, to Cavendish. Another 128 still lay ahead before it reached Burlington. On this particular day, Gage and his crew were working just 1,300 yards south of town, beneath the sheltering shadows of nearby Mount Ascutney, in a place known as Cavendish Gulf. It took at least three men most of the day to drill a twelve-inch hole in a granite outcropping, all in preparation to blow it apart. Two men took turns pounding the bit with heavy sledgehammers, while a third held the bit in place, turning it ever so slightly between swings.

Sometime after four o'clock, the men stopped hammering and Gage placed a cloth fuse in the narrow hole they'd spent hours making and pushed the material down as far as it could go with the tapered end of his tamping iron. Probably custom-made, Gage's tamping iron was 3 feet, 7 inches long and weighed 13½ pounds. At one end it was 1¼ inches thick and tapered to a quarter-inch at the other. After placing the fuse, Gage poured in the blasting powder. The next step was for another member of the crew to add sand. It was this protective layer that blunted the upward thrust from the explosion, the better to contain the blast and shatter the rock apart. Only after the sand was added, and Gage tamped it down, would he light the fuse.

Gage and his men performed this drilling ritual many times before, but on this particular day a momentary distraction caused Gage to turn his head, ever so slightly, to his right just as he reached for his iron. Thinking his assistant had already poured in the sand, Gage lifted the metal bar over the hole and pounded down. Before he had even fully turned his head back around, a spark from the iron lit the powder, burning his hands and arms up to the elbows and sending the 13-pound tamping iron, pointed end first, directly into his face. The iron rocketed through the zygomatic arch just under the left cheekbone, penetrated the orbital bone beneath Gage's left eye, and pushed the eye halfway out of its socket. It then smashed through the left frontal lobe of Gage's brain and exploded out of the top of his head, soaring through the air to land some twenty to twenty-five feet away, covered in blood and brains.

Gage was blown backward by the force of the explosion and laid on the ground, face to the sky. If he lost consciousness, it was only for a few moments. His hands and feet twitched, and fragments of bone and brain protruded from the bloody opening in his head, but within minutes he began to speak to his dumbfounded crew. They carried him gently over to an oxcart, normally used to haul heavy railroad equipment, and hurried into town. Gage sat upright the whole way with his back against the headboard and his feet stretched out toward the end of the cart. When they arrived at the tavern, where the railroad workers were all staying, Gage stood and walked unsteadily to the edge of the cart. With the help of his men, he stepped down and they guided him up the steps of the hotel and onto the porch. Nearby, the Reverend Joseph Freeman watched as Gage was gently lowered into a chair.

"What happened?" the minister asked.

"The blast went off when he was tamping it and the tamping iron passed through his head," said one of the men.

"That's impossible," Freeman replied.

On the veranda, hotel guests mingled with railroad workers, curious to know what had happened to the poor man, and Gage obliged them by telling them the astonishing story.

When Dr. Williams approached Gage, he asked where the iron had entered and the railroad foreman pointed to a 1½-inch-long slit in his cheek, which had been partially obscured by dirt and gunpowder.

"It went through the top of my head," Gage said.

"Sure it's so," said one of the crew. "The bar is lying in the road below, all blood and brains."

Gage then stood up, bent over, and vomited, and as he did so, about half a teacup of his own brain matter fell onto the porch. Within the hour, the town's senior doctor, Martyn Harlow, arrived at Adam's tavern. He had taken over the medical practice of Cavendish's previous physician just a few years earlier and, along with Williams, he treated a total population of about 1,500. Harlow was also young, just three years older than Gage and less than five years removed from Jefferson Medical College in Philadelphia, where he had specialized in anatomy.

Harlow asked a few of Gage's crew to carry him into the hotel and up to a room on the second floor, and before long the bed Gage was laid in was soaked with blood.

"I hope I'm not hurt too much," he said to Harlow, before vomiting several times in succession.

It is unlikely Harlow told his patient what he surely must have been thinking: that Gage could not have been hurt worse and still be alive. The patient continued to hemorrhage from his wounds and to vomit every ten to fifteen minutes. His pulse was 60 and regular, and he remained conscious as Harlow and Williams shaved his scalp, picked bits of loose brain matter from the opening in his head, and then removed three triangular-shaped pieces of bone that had shattered when the tamping iron exploded through the top of the skull.

While looking for more fractured bits of bone, Harlow carefully slid the index finger of his left hand through the yawning cavity at the top of Gage's head. He then slipped the index finger of his right hand through the opening below Gage's left eye. Unbelievably, his fingers nearly met—right where the gray tissue of Gage's left frontal lobe should have been. Being "unaccustomed to military surgery," Harlow later described the wounds as "truly terrific."

In the manner of reconstructing a puzzle, the two doctors fit the skull fragments together as best they could, then taped the scalp closed, leaving a small portion of it covered only by a dressing, the better to let the wound drain. Unbeknownst to the men, this decision, more than any other, probably saved Gage's life. If they had completely closed the opening, Gage's brain, which was quickly swelling with blood, would have begun to press down on his brain stem, shutting off his respiration and killing him.

In the late hours of the evening, Gage vomited less frequently but was slightly delirious. He told the doctors he didn't need to see his friends, since he would be back to work in a couple of days, but he correctly gave their names and where each lived. When word of Gage's accident reached cabinet-maker Thomas Winslow, the carpenter visited the gravely injured young man and quietly measured him for a coffin.

Early the following morning, Gage's face was swollen and his pulse more rapid, although he still spoke sensibly, even asking who was now the acting foreman of his crew. Forty-eight hours later, Gage was restless, his fever high and his speech incoherent. The wound in his head became infected and an abscess formed under his left eye, which still protruded slightly from its socket. Eventually the eye would be removed, but over the next couple of weeks, with liberal doses of calomel, rhubarb, and castor oil, he slowly improved. In early October, just three weeks after his accident, he was able to sit in a chair for five minutes. Five weeks after the accident, Harlow noted that Gage had improved enough to get out of bed "with little assistance," but was "very childish and wishes to go home to Lebanon, New Hampshire."

What little is known of the psychological changes that Gage underwent comes from Harlow's medical notes, which he submitted for publication in the *Boston Medical and Surgical Journal* a year later.

In those notes, made a month after Gage's accident, Harlow reported that his patient's desire to go home had become "uncontrollable by his friends," and that he continued to make arrangements to do so despite their protestations. Gage also went for a half-mile walk to the store on a cold, damp day without an overcoat and with thin boots, which left him wet and chilled on his return.

"I find him in bed, depressed and very irritable," Harlow wrote, although three days later Gage was well enough that he appeared "to be in a way of recovering if he can be controlled" from his impulse to get up and go home.

In December 1848, three months after the tamping iron shot through his brain, Gage returned to his parents' farm in Lebanon. Although barely able to work on the farm for the next six months, he was well enough to travel to Boston in November of 1849 and in January of 1850, where he was examined by Henry Jacob Bigelow, a professor of surgery at Harvard. Bigelow later wrote that he saw no pathological side effects from the brain injury, nor did he note any changes to Gage's speech, intellect, or reflexes. The strange case supported Bigelow's "whole brain" theory—that cerebral functions were not localized, but rather arose from the working of the entire brain—a view that was contrary to the phrenological trend at the time.

Early in the nineteenth century, Austrian physician Franz Joseph Gall proposed that character and personality traits could be identified by the shape and structure of people's skulls. In support of his own theory, Bigelow presented Gage's case to his medical students and Gage later appeared at a meeting of the Boston Society for Medical Improvement.

Gage's mother later claimed that her son also toured the major towns of New England to show people his wounds and his famous tamping iron, which he carried with him everywhere. At some point in 1850, he briefly joined P. T. Barnum's New York City freak show, called the American Museum. The following year, Gage found work in a livery stable and coach line run out of the Dartmouth Inn in Hanover, New Hampshire, just a few miles from his family's farm. He

worked in Hanover, home of Dartmouth College, for the next eighteen months before alighting for South America, where he helped an
American entrepreneur, James McGill, set up a stagecoach and livery
line in Valparaíso, Chile.

Not much is known of Gage's time in South America except that
the stagecoach route, which he must have driven often, was a rug-
ged one and required long hours riding over rough terrain. In June
of 1859, Gage returned to the United States, sailing to San Francisco
where his mother and married sister had moved. He arrived in the
rural California outpost seriously weakened from the long voyage,
and it was several months before he was able to go back to work, taking jobs at various farms south of the city. In February of 1860, after
plowing for a day, he sat down to dinner and suffered a serious epileptic seizure. More rocked his body in the days that followed, increasing
in regularity until finally on May 21, after enduring a grand mal seizure, Gage died. He was just thirty-seven years old. He had survived
a devastating brain injury in 1848 and then lived another eleven and
a half years, but to many people who knew him before the accident,
Phineas Gage was never the same man.

Harlow did not learn of his famous patient's death until five years
later and he then prevailed upon Gage's mother, Hannah, to have
her son's body exhumed and his skull, along with the tamping iron
that had been placed beside Gage in his coffin, sent back east to him.
In 1868, twenty years after the accident and eight years after Gage's
death, Harlow described Gage's psychological changes in a paper presented to the Massachusetts Medical Society:

> The equilibrium or balance, so to speak, between his intellectual
> faculties and animal propensities, seems to have been destroyed.
> He is fitful, irreverent, indulging at times in the grossest profanity
> (which was not previously his custom), manifesting little deference
> for his fellows, impatient of restraint, advice when it conflicts with
> his desires, at times pertinaciously obstinate, yet capricious and
> vacillating, devising many plans of future operations, which are no
> sooner arranged than they are abandoned . . . A child in his in-

tellectual capacity and manifestations, he has the animal passions of a strong man. Previous to his injury, although untrained in the schools, he possessed a well-balanced mind, and was looked upon by those who knew him as a shrewd, smart businessman, very energetic and persistent in executing all his plans of operation. In this regard his mind was radically changed, so decidedly that his friends and his acquaintances said he was "no longer Gage."

Harlow did not specify who these friends and acquaintances were or even whether their reports came to him as first-person accounts. The questions are considerable: Were the changes for the most part in the months after the injury? Did his erratic behavior settle down in later years? Or was it exacerbated by the succession of seizures in the months leading up to his death?

Since 1982, three studies, all using scanning technology, have attempted to reconstruct Gage's injuries and the path of destruction in his brain. The most recent, by two Boston researchers, digitally re-created the line of fracture in Gage's skull from just under the left cheekbone to just left of the median line and slightly in front of the bregma, the place on the top of the head where the skull plates meet. The scientists concluded that the damage was almost exclusively to the left frontal lobe.

To the doctors who attended Gage, the brain was mostly an organ of perception and sensation. They knew in general that the loss of movement or paralysis from stroke, for instance, could be traced to the brain, but they knew virtually nothing about the neural processes involved in thinking and feeling. Dax and Broca both offered evidence that the ability to speak and understand language was located in the left frontal lobe. But not until the 1870s did Scottish researcher David Ferrier declare, citing the Gage case, that personality changes can result from damage to the left prefrontal lobe.

In humans, the prefrontal lobes are located just behind the forehead and are the source of higher executive functioning, judgment, and reasoning. In the years following Gage's death, his case was often used by scientists to either defend or denigrate the theory of localization of brain function. The phrenologists, who claimed the shape and protuberances of the skull revealed character and intelligence, believed the changes in Gage's personality vividly demonstrated damage to what they called the "benevolence" and "veneration" regions of the brain. But the antiphrenologists believed Gage's normalcy—his apparently intact intelligence and speech and the lack of paralysis—was evidence of the nonlocalization of cerebral function.

World War I provided all the brutal evidence needed to resolve the matter, confirming that different parts of the brain do, in fact, control different functions. Reports emerged from battlefield hospitals of soldiers with brain wounds, specifically to their frontal lobes, who had become childish, impulsive, and unable to plan ahead. While it remains unclear if Gage's catastrophic wounds did the same to him, he was nonetheless able to function independently: he held down many jobs, worked long hours, and at least in South America toiled under difficult conditions in dangerous and unfamiliar territory. If Gage was, to his friends, "no longer Gage," something else remained—a Gage that lived, worked, communicated with friends and family, and died, much too young, in their midst. The question that can never be answered is "Who did Gage think he was?" Chances are he wasn't aware of his altered personality, or that others thought him different, and no correspondences of his that might point to an answer have survived.

The most pressing question for those who knew and loved Sarkin was if he would wake up. That was their urgent concern. The question of how much he might changed in his thinking or feeling seemed almost trivial by comparison. They would be there for him, in whatever condition he was in, when he woke up. They just wanted him to live.

Chapter 9

ABSOLVE ME

Am I dead?

The question bubbled up from deep inside Sarkin's brain. He tried to speak, but his words just seemed to bob along the edge of awareness, then drift away. Beached on the sheets of his hospital bed at Presbyterian University Hospital in Pittsburgh, he struggled to make sense of where he was and what had happened. He saw lights and he saw darkness, but he was unaware of almost everything else. Sometimes there were voices, distant, as if heard under water. In his mind he could see land, but he could not reach it. Time unspooled. He felt as if he was traveling backward, starting over.

He was six or seven and he was outside in an ice storm. The sleet hit him like knitting needles poking at his face, plunging deep into his skin. He wondered whether he was bleeding, if he would be scarred for life, if people would pretend not to look at him because his face was so disfigured. He would get used to it, he thought, the embarrassed looks and the sad expressions. He would grow up and get a job in a factory where he wouldn't have to talk to people, and he would live alone and bury his anger as deep as he could and only in the briefest of moments before sleep, when the maw of the abyss seemed to stretch from one end of his life to the other, would he think about everything he had lost.

Sarkin lingered for weeks in a semicoma, not quite asleep and not fully awake. Medical problems followed, one after the other: pneumonia, septicemia, a bleeding ulcer, a heart attack. Meanwhile, he drifted in a kind of twilight. The hours and days blurred together. He felt like the professor in H. G. Wells's *Time Machine,* the days whizzing by in an indistinguishable rush, the sun rising and setting in a fraction of a second, light and dark changing places with every blink of his eyes. As time passed, he could hear voices more distinctly and he wanted to speak, but tubes snaked all around him, into his mouth, up his nose, down his throat. Kim and Elaine came to the hospital every day. Jane, who was an editor at *Vanity Fair* magazine, flew in from New York City every weekend, and Richard was there as often as he could break away from his medical duties. The family essentially moved into the Holiday Inn that had just opened two blocks away, all of them dropping off every night into a deep, exhausted sleep.

When stress kept Kim awake, she watched *The Tonight Show* with Johnny Carson and *Late Night with David Letterman* with the baby tucked in beside her, not rising until around ten in the morning. The intensive care unit allowed visitors only once every two hours, for just fifteen minutes, so Kim and Elaine tag-teamed, spelling each other so one of them could take care of the baby. When Kim put Curtis to bed every evening, she held out Jon's photograph so that he could kiss his father good night.

Despite all of her anxiety about Jon, Kim enjoyed her walks to the hospital, past the field where the University of Pittsburgh football players practiced in the afternoons, past the forty-two-story limestone Cathedral of Learning, and the stately Greek columns of the Soldiers and Sailors Memorial. Fall had come early in 1989, and the old maple and oak, sweet gum and dogwood swaddled the town in red, orange, and yellow. *How can I feel so happy?* she would sometimes ask herself. She couldn't deny it, though. This was a beautiful place, and she seemed to draw strength from the crisp, cool air and the casual merriment of the college students. She was comforted, too, by the thought that this was where she was supposed to be—with her husband. She couldn't imagine being home in Gloucester, without him. On those walks to and from the hospital, she also couldn't help thinking about the accidents of life—all those lessons we learn that we think we'll never need, until one day your husband comes home and says something happened in his head, and nothing is ever the same.

Kim had an inkling of some of those lessons when she was a student in Ecuador her junior year in college. She learned then how fragile life is, and how destructible our plans for the future. At the time, she lived with several other women in a gated apartment complex with guards in front of the building. Next door to her building was an empty lot where several families lived in haphazardly built shacks. On the other side of the lot was a mansion. The inequities were deeply upsetting. On weekends, the student teachers would pile into a Jeep and head out into the country. One Sunday they pulled up to a gorgeous cathedral as people were streaming up the stone stairs and through its large front doors. As they did, they passed by and stepped over a destitute man lying on his side. Barely clothed, his cheeks sunken into his skull, he held out his hand, begging. The scene horrified Kim, who realized the man was probably dying.

A few months later, in February 1981, Kim and a few friends were again riding in an open Jeep in an unfamiliar section of town. As they turned a corner, they suddenly saw a pile of tires burning in the middle of the street. Tear gas began to sting their eyes. They quickly turned around and headed back to their apartment. That night their sponsors told them they needed to stay indoors, sleep in their clothes, and have their passports and airline tickets at their bedside in case they needed to be evacuated in a hurry. Ecuadorian and Peruvian armies were clashing over a disputed border region east of the Condor Mountains that had been a source of conflict for at least forty years. The battles raged for two months, and the experience of coming so close to being swept up in the conflict left Kim with a deep impression of the mercurial quality of fate.

As focused as she was on Jon's recovery, she was grateful for small breaks from the hospital. Still, her focus on Jon, her connection to him even when she wasn't with him, was powerful. Once, during the long weeks of Jon's recovery, Kim's father flew in from the East for a few days. He took his daughter and baby grandson to the Pittsburgh zoo, which was just seven miles away from the hospital, but Kim was uncomfortable the whole time. She felt sick to her stomach being even this far away from Jon.

"I have to go back," she said to her father, calling the day short.

Jon usually opened his eyes when she was there. He would look at her, squeeze her hand or nod his head, then fade away again into that in-between place where he always seemed just about to wake up.

When Jon was moved out of the ICU to a step-down unit in October, Kim took the baby to see him, thinking it might cheer him up to be with his son again. Jon was still intubated and couldn't speak, but he was more alert, more responsive. When Kim brought Curtis over to the bed, Jon became agitated and motioned for her to leave. He didn't want Curtis to see him like this, with all the tubes and wires running in and out of his body like Frankenstein's monster.

The doctors had told him there was considerable damage to his cerebellum because of the blood clot, and that much of the left side of the brain structure had been removed. Unlike the rest of the cerebrum, the left cerebellum controls the left side of the body, and vice versa. The result, they said, was that he was deaf in his left ear, weaker on his left side, and would always have trouble with his balance, his speech, and his vision.

In October of 1989, Sarkin was flown by medical helicopter from Pittsburgh to Boston, then driven to the New England Rehabilitation Hospital in Woburn, Massachusetts, about thirty-five miles south of Gloucester. Within a week, his condition spiraled precipitously downward. An IV line in his neck had caused an infection to spread throughout his body. With a fever of 105 degrees, Sarkin was moved to nearby Beverly Hospital, loaded with antibiotics, and placed on a bed of ice to bring his temperature down.

"How's he doing?" Kim asked one day when she walked into the ICU.

"If he makes it through this . . . ," the nurse started to answer.

Kim didn't hear the rest of the sentence. *If he makes it through this?!* She hadn't realized just how close he was to dying again. The whole life-and-death part—she thought that was all in the past. Now they were back in the trenches, waiting again for new signs of hope.

When John Keegan visited, he was stunned by his friend's condition and began to cry. He hadn't seen him since their poker game two months earlier.

"I'm sorry, Jon. I'm so sorry," he whispered.

Sarkin stirred, raised an emaciated arm toward Keegan, then rubbed his thumb against his other fingers—the universal sign for "hand over the money." Keegan couldn't believe it. The guy in the bed looked nothing like his friend, but he remembered the poker game and the money Keegan still owed him. As bad as it all was, the real Jon still seemed to be in there.

A few days later, Sarkin swam up out of the darkness yet again. No longer intubated, he raised his head off the pillow when Kim entered the room. Slowly, painfully, he tried to speak for the first time in more than two months.

"Kim?"

"Oh my God, you can talk!"

He tried to put the words together.

"Absolve me," he whispered to his wife. "Absolve me."

"What are you talking about, Jon? For what?"

"For my sins."

Kim was confused. She'd never heard Jon speak like this, using religious language. He wasn't even Christian. She walked out of the room and found a nurse.

"He's talking, but I don't understand what he's saying. 'Absolve me of my sins?' What does that mean?"

"Oh, all patients say crazy things when they're in the ICU," the nurse said.

In truth, there was nothing to forgive, because nothing had been Jon's fault. What happened during the microvascular decompression to cause the hemorrhage would remain a mystery. Jannetta viewed the videotape of the operation and read over his surgical notes, trying to figure it out, but to no avail. When the family considered a lawsuit, a team of independent neurologists studied the tape and found that Dr. Jannetta had made no obvious mistake. In medical parlance, Sarkin's stroke had been an "unforeseen consequence." In short, an act of God.

When he returned to the rehabilitation hospital in Woburn in November of 1989, progress was slow and tortuous. He took weeks to learn how to breathe, eat, and speak again normally. When he did, one of his first phone calls was to his sister, Jane. One day in the late fall, she found a single sentence from an assistant scratched on the message pad on her desk:

Your brother Jon called.

Jane cried when she read it. She'd never really believed her brother would die, but after all these months of pain and waiting, this was like a message from heaven. She tucked the note away in a safe place, as a keepsake, hoping it was a sure sign of Jon's return to the world.

By far the most agonizing chore for Sarkin at the beginning of his recovery was simply learning to sit. Trying to maintain his balance in a chair was like riding the tilt-a-whirl at an amusement park—everything moved. As the room spun around mercilessly, it was all he could do not to vomit. The problem was that his damaged cerebellum was sending distorted signals to his ears, upsetting his balance and equilibrium.

Normally, the structures of the inner ear enable the body to adjust by sending electrical messages through the vestibulocochlear nerve and then finally into the processing center of the cerebellum. With half his cerebellum missing or dead, however, Sarkin's brain couldn't make sense of the messages from his ears. When he sat up he swirled, when he tried to stand he tumbled. All he wanted to do was lie down, but his physical therapist insisted he sit. The orderlies picked him up out of bed and deposited him in a chair, strapping his arms and legs down so that he wouldn't try to stand. Sitting was so nauseating, though, that one time he tried to stand anyway in order to get back into bed, and fell flat on his face, bruising his head and bloodying his nose. His weight plummeted from 170 pounds to 145. Infections and fevers wracked his body and each bout seemed to leave him a little less of the man he was before.

The cerebellum controls many of the muscles used in fine-motor coordination, and not just the muscles in our legs, arms, and fingers, but in our faces. The impairment to the muscles that allow a person to speak meant that Sarkin's speech would always be slightly slurred, and the crippling of the muscles that allow the eyes to track in tandem meant he would see double the rest of his life. Following moving objects would always be difficult, catching a ball challenging, hammering a nail dangerous, and riding a bicycle impossible.

At least 85 percent of our perception, learning, and cognition is mediated through our eyes. The stroke disrupted several visual centers in Sarkin's brain, which meant that instead of a constant, seamless flow of information from the world, images stuttered through his brain like the frames of an old 16 mm film when it tears and tumbles off its reel. He also suffered from a condition called nystagmus, a jerky motion of the eyes, which often gave him nausea, a sensation of vertigo, and the feeling that the world was in constant motion. Sarkin would never be able to run, climb a mountain, or play golf again, and he would always need a cane. Lying in his hospital bed, or sitting in a chair, which he loathed, he looked cadaverous.

The extent of his physical damage dawned slowly on Sarkin. One evening when he was still in the rehabilitation hospital, he tried to call Kim at home, but when he raised the receiver to his left ear, expecting a dial tone, he heard nothing. Maybe the phone was broken, he thought. He jiggled the receiver, pressed the button down several times, and waited. Three, four, five seconds. Still nothing. Confused and frustrated, he returned the receiver to its cradle—and then he remembered. He had held the receiver up to his left ear, the one with which he could no longer hear.

For the present, all Kim could do was comfort her husband. They would get through this, she told him. But when she looked into his eyes now, she didn't quite recognize him. He seemed slightly wild-eyed, as if only loosely tethered to the present and to her. The man she'd married had been both fun and serious. He had jumped in with the band at their wedding reception and played a few rock solos on the electric guitar, but he was so grounded, too, so dependable: never late with a bill, always putting money away. He had been thriving in his practice before the tinnitus, before the stroke, before the surgeries and everything in between. Now he seemed self-absorbed, uninterested in her or Curtis, and listless.

One day Kim brought Jon his guitar to the rehab center, hoping it might offer a good distraction and a welcome reminder of life back home. He didn't touch it while she was there, but when she left, he picked it up and felt its weight in his lap. He stroked the strings with his right hand and then searched for a familiar chord with his left. His fingers were stiff, uncooperative, and foreign, as if they belonged to a hand that wasn't his. He couldn't coordinate the left with the right and nothing he tried seem to work. He put the guitar down and never picked it up again.

One of the first friends to visit him in rehab was Hank Turgeon. He was the first person, outside of Sarkin's family, to learn about the stroke when he called to check on Jon the night of his first surgery. Now, all these months later, he was walking into a rehab center in Woburn. Sarkin had been briefly rehospitalized for pneumonia and was back on a breathing tube. Turgeon checked with the nurse at the front desk and was directed to a room on the third floor, but when he walked in and saw a haggard old man, he thought he must be in the wrong place, turned, and walked out.

"Where's Jon Sarkin?" he asked the nurse at the desk again.

"No, no, that's him," she said, pointing to the room with the old man. "That's your friend."

What have they done to this guy? Turgeon thought. Sarkin was propped up by pillows. His skin was the color of pewter, and he was unable to speak. Turgeon walked over to the bed, sat down, and squeezed his friend's hand.

Weeks later, when Sarkin no longer needed breathing assistance, Turgeon brought his guitar, thinking he might try to cheer his friend by playing a little music. When they had first met several years earlier, he had been blown away by Sarkin's musicianship, the way he could play the slide guitar and riff on Robert Johnson's "Stop Breakin' Down Blues," or play "Route 66" the way Chuck Berry did.

"What do you want to hear?" Turgeon asked

" 'Tears of Rage,' " Sarkin answered.

The Bob Dylan song, an anthem from the late 1960s, was moody and emotional. Despite the months of having a tube down his throat, Sarkin joined in and Turgeon was amazed. They sang every verse of the song, with Sarkin not so much singing as spitting the words out in a hoarse, angry voice. When they finished, Turgeon looked up.

"What do you want me to play now?"

" 'Desolation Row.' "

Another Dylan song. Sarkin was half-lying, half-sitting up in bed. A nurse came in, stared at the two men, then turned and left. They went through all eight verses, Sarkin keeping tune, half-singing, half-speaking.

Yes, I received your letter yesterday
(About the time the doorknob broke)
When you asked how I was doing
Was that some kind of joke?
All these people that you mention
Yes, I know them, they're quite lame
I had to rearrange their faces
And give them all another name
Right now I can't read too good
Don't send me no letters, no
Not unless you mail them
From Desolation Row.

Turgeon was chilled by his friend's song selections. This was a different man. A bitter man. He couldn't exactly blame him, but there was something foreign now about Jon. He was moody and angry and—this was hard for Turgeon to admit—a little unrecognizable. He never discussed any of this with Kim, of course, but he needn't have. She had seen with stark clarity how different Jon had become. He seemed unable to hold anything back, as if the speed governor on his car had been smashed and there was no limit now to what he might say. If he didn't like what someone was wearing, he said so; if he thought someone was boring, he told them.

Sometimes, when Jon was in physical therapy, Kim stayed to watch. Once, she sat nearby as a young female therapist guided him through the parallel bars, encouraging him, one small step at a time. Suddenly, without warning, he blurted out that he wanted to have sex with her. Kim's eyes widened in horror. One of the things she'd always liked about Jon was that he was very proper in an old-fashioned kind of way. He had never been the type to ogle other women, and she certainly had never felt an ounce of jealousy or the need to worry about him wandering. Intellectually, she knew this outburst of his was the result of the stroke, but part of her couldn't help but feel hurt. She ran from the room in tears. When she got home, she called the hospital.

"Please tell him I'm not mad at him," she told the nurse. "Please just tell him I had to leave."

The words had come bursting out of Sarkin. No hesitation. No premeditation. Disinhibition often occurs in patients with injuries to their frontal lobes, the part of the brain responsible for social behavior. Usually the behavior goes away, but how long it takes is anyone's guess.

Kim knew Jon's body was now vastly different. His eyes were crooked, his speech was slow, he was deaf in one ear and weak on one side, but no one—doctor, nurse, or social worker—had ever talked to her about changes in personality or behavior. No one had told her he might not think or act the same, that he might actually seem like a different man. All they said was, except for the balance and hearing, he would most likely make a complete recovery. She had been so intent on Jon's physical progress, on getting him to walk and talk. Now another thought began to nag at her: Would there be other surprises, other changes that would make her feel she did not know who her husband was anymore?

Chapter 10

PUZZLE PIECES

S arkin returned home in a wheelchair in December 1989, four months after he'd left Gloucester for Pittsburgh, hoping a risky operation would restore him to health. Virtually everything that could go wrong had, but he'd survived. His life was still pared down to the basics—eating, drinking, and sleeping—but Kim had him home, which felt like a huge step back toward normalcy. He quickly moved from the wheelchair to a walker, but it would take another six months to transition to the cane, and he would depend on it the rest of his life. Walking would never feel normal. A lingering sense of vertigo meant no matter where he was, he would always feel like he was just about to drop off a cliff. All of his senses were skewed. He spoke too loudly, probably because of his hearing loss, and colors seemed to jump out at him, maybe because of his trouble distinguishing background from foreground. Even food smelled and tasted different.

The left part of his cerebellum had been removed, which accounted for many of those physical changes, but there was also damage elsewhere due to lack of oxygen when his heart stopped twice in the hospital. Even though neurons had shriveled and died all over his cortex, however, he retained the ability to speak, reason, read, write, and remember. The stroke had, in effect, tiptoed across his brain, sparing the most important functions but damaging a host of others, including his ability to focus and plan, to think sequentially, and act predictably. Thinking things through—whether it was getting dressed in the morning or reading the newspaper—was almost impossible. Instead, his mind rambled; he turned to do something, and something else would catch his attention. Nothing was ever finished: his hair was half-combed, meals turned cold, conversations were left hanging.

The most unsettling part for Sarkin was the sense that there were no filters, no chance for his brain to slow everything down and order the world into meaningful images and scenes. And combined with his inability to focus or filter, his double vision and lack of depth perception blurred the world. Everything overlapped. He was like a man watching a parade, unable to make out the music of the band passing by because it was already mingling with the tune of the one approaching. Every minute of his life was a cacophony, with each jarring sensation competing for his attention. The world didn't flow or float by Sarkin, it stutter-stepped, and he felt constantly out of synch. And because nothing was filtered, everything was equally interesting. Nothing escaped his attention, even if nothing held it, either.

In the mornings, he sat in front of the TV with two-year-old Curtis and laughed crazily at Mickey Mouse cartoons. He and Kim talked, but never really conversed, not like husband and wife. Kim would ask Jon what he needed: Was he hungry? Did he need help getting dressed, going to the bathroom? She encouraged him, telling him he was going to get better, that they would get through this together, reminding him that the doctors had said he would make a nearly complete recovery.

Still, Kim knew Jon had suffered intolerably. He had nearly died several times. So, in a way, she wasn't surprised when he started exhibiting signs of obsessive behavior that centered on his determination to prepare the family in case of future disaster.

"We need three things," he said urgently to Kim one day. "Progresso kidney beans! Progresso lentil soup! Progresso minestrone!"

"Jon, that's not really necessary," she said.

"We need them!"

"All right, Jon, how many cans?" Kim asked, trying to be patient.

"Not cans! We need cases. At least ten cases!"

Then he told her they needed to store jugs of spring water, gallons of gasoline, and cans of Sterno in the basement.

"Maybe we should buy gold bars and bury them under Curtis's sandbox?" he suggested another day.

He wanted to save electricity, too, so they often lit candles instead of turning on the lights at night. And everything—*everything*—had to be recycled.

"But, Jon, there's no recycling here."

"No problem, we'll send it to my brother. They recycle in Buffalo."

Kim diligently bundled up all the things her husband deemed recyclable—everything from paper napkins to ballpoint pens—and stored them in the basement. She told Jon she would mail them to Richard, but instead of going to the post office, drove the bags to the town dump.

The recycling extended even to food. The entire Sarkin family—Jon, Kim, his mother, stepfather, brother, sister, their kids and spouses—took a vacation in Florida when Jon was finally out of the wheelchair. Every day they went to the beach, and every day someone took a lunch order and brought back meals from McDonald's. Sarkin's choice was usually chicken fingers and French fries. When he couldn't finish the fries, he tucked the half-eaten bags into his backpack. By the end of vacation, oily bags of French fries had piled up in the hotel room. Kim didn't challenge him about it. There were too many other problems—getting Jon and the baby dressed in the morning, taking them to the beach and back every day—to linger over every one of his eccentricities. Most of the time she tried to ride out the waves of bizarre behavior, hoping that in time the waves would subside and the old Jon would reemerge. Only when something became a problem—

an issue of sanitation or safety—would she question him. She never left Jon alone with the baby, not because she thought he might hurt Curtis, but because she just couldn't trust him to keep an eye on him. God forbid something should happen; she wasn't even sure Jon would know what to do in an emergency.

That day leaving the beach, when he started packing French fries in his suitcase, she finally stopped him.

"Jon, what are you doing?" she asked.

"I'm going to recycle them."

"Recycle them where?" Kim asked.

"Home," he replied, as if that was a perfectly normal thing to do.

She convinced him that the French fries could just as easily be recycled in Florida as in Massachusetts, and that she would take care of it, later throwing them away when he wasn't paying attention.

His mind seemed to wander incessantly, and full comprehension was always just outside his reach. In the middle of a thought something else would hijack his focus until all he was left with was an idea or an image that was almost complete, but not quite. Everything in his mind was just slightly out of place, a little like those model car parts he assembled as a kid with the wheels glued to the roof. He had no idea how to put the puzzle of his life back together. Complicating the reassembly was that everything was out of proportion—his emotions, perceptions, what was important, what was not. He was like a funhouse reflection of himself—all the parts were there, but distorted—the body was too long, the head too wide, the features of his face smeared like putty.

Months after returning home, Sarkin was still receiving get well cards and phone calls from friends inquiring about his health. One day he suddenly decided to respond.

"Kim, I need all their addresses," he said.

"Why?" she asked.

"I have to send thank-you notes."

"Jon, that's not really necessary," she said. "People who send get well cards don't expect you to send a card in return."

For Kim, it was just one more bit of evidence that as much progress as Jon had made walking and talking, he was still far from back to normal. She understood where the card-writing impulse came from. They had been overwhelmed by other people's good wishes—cards, letters, phone calls—when they returned home from Pittsburgh, and Jon's instinct was to thank them all. He was a polite, do-the-right-kind-of-thing guy. That was one of the qualities she loved about him. But now he lacked perspective. To her relief she convinced him there was no way he could respond to everyone, and that in any case no one expected him to.

Periodically, Kim got more frightening indications of just how seriously impaired Jon had become. She drove him to physical therapy three times a week, and on one occasion, not long after he'd returned home, he rode in the backseat with Curtis. Distracted, Kim had forgotten to buckle the baby in, and as they were driving down the highway, Curtis crawled out of his car seat and squeezed through the gap between the driver and passenger's seat. Suddenly, he was standing next to Kim, while Jon, still in the back, was oblivious.

"Oh my God, Jon, Curtis is out of his car seat!"

She quickly pulled over to the side of the road, got out with Curtis in her arms, opened the back door, and buckled him back into his seat.

"Didn't you see him?" she asked, exasperated. "Why didn't you tell me he was out?"

Sarkin just shrugged. He had no answer; it just hadn't occurred to him. For Kim, this was what exasperated her most—not his physical incapacities, but this mental and emotional fog. He existed *in* the family, but he wasn't *part* of it, not really, not like normal husbands and fathers. He didn't seem to feel a normal kind of remorse, either, or want to atone for the things he'd done wrong. Instead, his contrition took the form of self-absorption, harshly castigating himself, so that Kim ended up not only caring for Curtis but comforting Jon.

Friends and family had a difficult time understanding just how hard it was for her, because they didn't see the full extent of his impairment. When his mother, sister, or brother visited, each focused on how much better he was doing physically. His strange behavior was harder to perceive since they didn't live with him. Not long after Jon came home, a friend of the family suggested to Kim that maybe he should try aromatherapy. She was stunned, suddenly realizing how little other people understood about strokes and just how profoundly they can affect a person's behavior.

"They removed a part of his *brain*!" she said to the woman.

At night, with Curtis curled up tight against her chest and Jon already asleep beside her, Kim would think about the future, about all the years yet to unwind ahead of them, and like a soft but steady snowfall, a sense of longing and loneliness enveloped her. In a way, she had been hibernating for months. Since the stroke, she had co-cooned herself inside her own mind, unable to share her deepest fears and thoughts with Jon. What if this was as much recovery as he'd ever be capable of? The prospect that *this* Jon, the poststroke one, so distracted, eccentric, and often depressed, was the one she would have to live with for the rest of her life, was not something she was willing to face yet.

Kim confided her fears and concerns to her mother, sometimes her brothers, and a few close friends, but her main refuge was Curtis. Before the stroke, before the tinnitus, she and Jon had shared equally in the bliss that Curtis's birth had brought them: sleeping together, taking walks together. Lying on the bed or on the couch, Jon would toss Curtis in the air until the baby dissolved into fits of giggles. He couldn't do that now, and even worse, he didn't seem to want to. There were times when Kim felt robbed, robbed of her future and of the life she and Jon had just been starting to build together, and she mourned for her own loss as much as his. Occasionally Kim's friends would ask her, gently, how she put up with it all.

"Don't you ever think about leaving?" they asked.

"Sometimes," she would say, "but not really."

Kim had liked being a traditional mother and wife, not having to work and feeling as though she could spend all her time on child rearing. She still didn't need to worry about working. Luckily, Jon had taken out a sizeable amount of private disability insurance when he opened his practice, and combined with the disability income from the federal government as well as money he had invested, they would always be able to live a good, if frugal, life. But now she was in charge not only of the baby's welfare and being a homemaker, but of the finances, the driving, and even household repairs. In their brief time together before the stroke, she had liked being able to depend on Jon to keep her safe and secure. Now she could rely only on herself, and she didn't know how long that would be the case. It might be forever. If someone broke into the house while they were asleep, she wondered if Jon would have any idea what to do.

Sometimes Sarkin tried to be helpful, but there was little he could reliably do. He surprised Kim one day by putting away the groceries and cleaning up the kitchen, only everything was in the wrong place: the meat in the pantry, the milk in the cupboard, the toilet paper under the sink. Any activity was a complex action for him, a puzzle he had to solve, and more often than not when he completed it, the pieces didn't fit. One time he filled the dishwasher with the dirty dinner plates, pots, and pans, but instead of detergent, filled the cup with shampoo. Several inches of suds and water flooded the kitchen floor.

The thought of leaving him, though, was unacceptable to Kim. He depended on her now, and in any case, being a single mother was out of the question. Kim's parents divorced when she was thirteen, so she knew how difficult that road was, not only for the parents, but especially for the children. She couldn't imagine doing that to Curtis—or herself. In any case, the vows she and Jon took when they married meant something to her. People gave up too quickly on their relationships, she thought. And dating: Did she really want to do *that* again? Jon had given her everything she'd ever wanted, a husband, a home, a child. He no longer could provide the emotional and physical security she needed, but growing up with three brothers had taught her one thing—how to tough it out.

She also wasn't willing to give up on her dream of having more children. All during Jon's illness and recovery, she had not stopped thinking about getting pregnant again. Another baby might be healing for all of them, she thought, might remind Jon of those nine months of wonder they had with Curtis and help him return to his former self. She knew it was probably selfish, since Jon was still so challenged, but she didn't want Curtis to grow up an only child without the companionship of either a healthy father or a sibling close in age.

Having another baby would be healing for her, too. Motherhood was the most important, most rewarding part of her life. In the early months of 1990, she began to talk to Jon about getting pregnant again, reminding him of their original plans. Now that Curtis was two, maybe it was the right time, she said. Jon didn't fight her. Even if he couldn't fully understand the responsibilities ahead, Kim thought it might help restore in him a sense of their old life together, and offer proof to them both that they were somehow still a family. They agreed to start trying.

Chapter 11

ADRIFT

As the winter of 1990 turned to spring, Sarkin seemed stuck. If he'd awakened from his stroke more like an infant, having to learn the basics of life again, he was now more or less like a teenager. There were occasional moments of exuberance, but mostly he was sullen and withdrawn, and Kim had to keep encouraging him. Usually they just wandered around each other, like two planets whose orbits rarely intersect. For her part, Kim was almost totally consumed with caring for Curtis and doing all the household chores, paying the bills, and balancing the checkbook.

Occasionally Jon changed the baby, a task made slightly more difficult because Kim used cloth diapers. When she checked, she often found the diaper askew so that the pins were rubbing against Curtis's skin, irritating his belly, so she would redo it. One thing Jon could help with, though, was picking up the playroom. He would sit on the floor and slowly, methodically, put toys back on the shelves. When she thanked him, he told her it gave him pleasure.

"I decided this should be my job because I can do it," he said.

He also began to write things down, fragments of thoughts, words that popped into his head. Occasionally he doodled, sometimes just circles or spirals or rows of zigzagging lines. Other times he drew cartoon faces assaulted by knives, riven by pipelike instruments, or fractured and distorted—hallucinogenic recollections of the deathly ill man lying in a hospital bed with tubes running in and out of his body.

In the summer, he picked up his old guitar case, which had been standing in a corner, untouched, since his aborted attempt to play in the hospital. Now, instead of taking the instrument out, he placed the case flat on his lap and drew on it. First, he covered it with white and blue pastel, then added a green cactus, then another cactus and another, until the entire backside of the guitar case was filled with nine spiky plants marching across a desert landscape, framed by a pale blue sky.

Gradually, Sarkin became more intent about his drawing; there was an urgency to it, as if he was filled with a need to get each image, each word, each thought out of his head, and so he often drew the same word or image again and again. His odd emotional detachment and lack of awareness also continued. If he was at the dinner table with Kim and a picture popped into his head or a song lyric, he would get up in the middle of the conversation, with no warning, and find one of his drawing pads. Sometimes Kim would chastise him for being rude, and only then would he realize he'd done something wrong. Embarrassment, shame, guilt just didn't seem to register with him anymore. He didn't mean to hurt or ignore Kim; the problem was that he was conscious only of the moment, able to discern only the "now" of his life.

The odd way he often acted was symptomatic of specific damage to Sarkin's brain. Perseveration, or compulsive repetition, such as his nearly obsessive drawing of cactuses, was an indication of frontal lobe damage. For Kim, however, the most disturbing changes were Jon's self-absorption, his insensitivity toward others, and his apparent lack of empathy. The source of these changes certainly involved the frontal lobes, but more particularly the damaged mirror neurons in the frontal lobes, a brain system scientists have only recently begun to understand.

The word *empathy* was first introduced into the American lexicon by psychologist Edward Titchener, a popular teacher at Cornell University from 1892 until his death in 1927. Titchener's "empathy" was a translation of the German *Einfühlung*, or "in-feeling," a term coined by a German philosopher of aesthetics, Robert Vischer, in 1872 to characterize the feelings a piece of art evokes in a viewer. Another German, Theodor Lipps, used the word to describe the feelings of emotional, even physical connection we have when, for example, we watch an acrobat on a high wire and our hands perspire, our stomachs feel queasy, and we tense our bodies in concert with the circus performer, bending this way and that as he struggles to maintain his balance. Empathy, said Lipps, was a natural by-product of being human, and being human, we are prone to putting ourselves in other people's shoes. Not until the latter half of the twentieth century, however, was empathy analyzed scientifically and, as it turns out, accidentally.

On a blazingly hot Italian summer afternoon in 1991, researcher Giacomo Rizzolatti was in his lab at the University of Parma investigating voluntary motor action. He and his colleagues were using macaque monkeys, a primate close to humans on the evolutionary ladder, to see what happened in their brains when they reached for a seed or a raisin. Throughout this particular morning, the electrodes implanted in the premotor cortex of a particular macaque sparked with activity whenever the primate reached for a piece of food on a tray and brought the morsel up to its mouth.

As one of Rizzolatti's graduate students set up a second experiment, the neuroscientist noticed something peculiar. Although the macaque was sitting quietly in its specially devised chair, the machine tracking the neurons in its premotor cortex was crackling, and in exactly the same pattern as if the macaque were reaching for a piece of food, only it wasn't. However, the macaque *was* watching a graduate student prepare the next experiment, reaching out and placing various seeds and raisins on the tray. What was happening? Rizzolatti wondered. Had the macaque, unseen by anyone, moved ever so slightly toward the tray and set off overly sensitive electrodes? While he pondered the strange response, several other graduate students trickled back into the lab from their lunch break. One of them had an ice cream cone, and as he raised the cone to his mouth, once again the electrodes attached to the macaque's premotor cortex started to chirp, and again in the same pattern as when the primate had grasped a piece of food and lifted it toward its mouth.

Rizzolatti was more than intrigued. For the next few years, he continued to do experiments that replicated what he'd seen in the lab that day. Consistently, he was able to provoke premotor cortex signals in the brains of the macaques while they were watching an activity, not performing it themselves. Eventually, Rizzolatti postulated an entirely new brain function connected with what he called "mirror" neurons. Nerve cells in the premotor cortex, he said, are activated not just when a macaque moves, but when it watches another animal—or human—move in a similar manner.

Since Rizzolatti's accidental discovery, hundreds of studies of mirror neurons have been conducted both with animals and humans. Research has located these specialized cells in a wide range of brain areas, not just the premotor cortex. In 2005, Rizzolatti and colleagues found mirror neurons in the language-processing part of the brain, discovering that they fired when subjects merely listened to someone describing an action. Similar research by UCLA's Marco Iacoboni discovered that expectation, or the perception of intention in others, is also linked to mirror neurons. In one experiment, subjects watched a film of different tea-related scenes. One of the scenes appeared to be taking place at the start of tea time, another at the end, and researchers found that entirely different sets of mirror neurons fired in the brains of subjects looking at the scenes depending on whether the person in the scene appeared to be reaching for the teacup to drink it or to clean it. The mirror neuron system, they concluded, was incredibly complex and involved not merely in empathy, but in many other specific and nuanced cognitive processes.

Mirror neurons are associated with a whole range of mostly social traits. They light up when a person observes emotions, from disgust to joy, in another's face. They are activated when someone hears a joke, expresses thanks to someone, or feels ignored. To many people, this comes as no surprise. After all, when someone winces in pain we wince, too. When we see a person slam a finger in a car door, we reflexively grab our own. Even when we're just being told about an accident or injury, we cringe and grimace in horror.

In the 1990s, two New York University researchers, Tanya Chartrand and John Bargh, conducted experiments that showed that merely observing another person's behavior increases the likelihood that the person doing the observing will unconsciously engage in that behavior as well, a phenomenon they dubbed the "chameleon effect." Scientists have found that speech is particularly susceptible to imitation in tone, rate, rhythm, syntax, even choice of words, which is why the more time we spend with someone who has a distinctive accent, the more we begin to sound like that person. (Those who have criticized pop singer Madonna for her pseudo-British accent should take heed. She was living in England, so she couldn't help it.) Yawning in response to someone else doing it is similarly unconscious, hence the term "contagious yawning."

The unconscious, even involuntary nature of imitative behavior suggests that it is also primitive behavior, and therefore that mirror neurons may have played an important role in human evolution. Certainly, imitation, or the imaginative imitation of another person's actions, is instructional. When we watch someone play an unfamiliar game, we mentally rehearse the moves of the player we're watching. We learn how others act by observing and reenacting. Learning how others think and feel may have been just as important.

Dutch scientists took the study of the chameleon effect one step further in 2002 when they created an experiment that showed if a waitress simply mimicked, or repeated, the order given by a customer, her tips increased. Why we have mirror neurons has not yet been fully answered, but the Dutch experiments suggest that imitative behavior works as a bonding device, and bonding or cooperative behavior has been proven to be evolutionarily adaptive.

The reason we don't all walk around repeating each other's movements, however, is that most of the time, especially in social situations, neurons that inhibit movement override the impulse of the mirror neurons to "act out." Not surprisingly, damaged or disabled mirror neuron systems can result in disorders such as echopraxia, in which a person experiences a powerful compulsion to imitate the actions of anyone with whom they come in contact. Rarer still are cases of extreme empathy in which simply by observing someone else's pain, a person appears to experience the same physical pain in their own body.

In 2003, as British neuroscientist Sarah-Jayne Blakemore was giving a talk in which she mentioned anecdotal stories about people who feel "empathic pain," a woman in the audience raised her hand.

"Doesn't everyone experience that?" she asked Blakemore. "Isn't that completely normal?"

The neuroscientist eventually met and tested the woman and discovered that when she faced someone who was touched on the right cheek, she experienced the same sensation on her left cheek. When the experiment was repeated with the two standing side by side, the woman felt the touch on the same cheek as the woman performing the action standing next to her. The empathic touch was experienced exactly as it was observed.

Blakemore calls this unusual phenomenon "mirror-touch synaesthesia." Synaesthesia usually refers to a condition in which one type of stimulation evokes the sensation of another, like a sound that has color or a smell that is tactile. People who have mirror-touch synaesthesia, says Blakemore, have an overactive mirror neuron system. It's not the skin or muscles or bones that register sensation. It's the brain. Therefore, she says, it is entirely conceivable that an extremely high level of empathy caused by overexcited mirror neurons triggers pain sensors in those rare individuals who can't help placing themselves in other people's shoes.

At the opposite end of the spectrum are those who because of disease, disorder, or an injury to their brain are unable to empathize at all. V. S. Ramachandran, director of the Center for Brain and Cognition at the University of California–San Diego, believes that damaged or deficient mirror neurons may be one of the reasons why so many autistic individuals are unable to engage with others socially or emotionally. Using several brain-imaging devices, Ramachandran tested subjects with high-functioning autism and found that their mirror neurons responded only when they were performing an action, not when they were observing it in someone else.

The ability to understand other people's beliefs, intentions, and desires is called having a "theory of mind," coined by psychologists in 1978, and it is a critical cognitive tool in human development. Children develop a full theory of mind by about the age of four, but it begins at just a few months of age when an infant learns to differentiate between facial expressions. Scientists such as Ramachandran now believe that certain aspects of autism can be explained by difficulties in a person's theory of mind, and problems in a theory of mind, he believes, are inevitably linked to disorders of mirror neurons.

With Sarkin, the absence of interest in others was effectively a kind of alienation and sometimes made him seem slightly autistic. With his physical recovery having progressed as far as it possibly could, he was becoming more aware of a vacant feeling inside himself and more upset by it. The map for reclaiming his body—walking, talking, breathing again—had been laid out clearly, the steps followed, the progress made and measured. But the stroke had also dislodged something just as crucial to his life—his sense of himself. He remembered who he had been before his stroke, but at the same time he knew that he was different now; he was estranged from himself. How would he ever be able to put the pieces back together when he wasn't even sure what he was missing? He was adrift, unanchored to time or place or person, and unsure if he would ever find safe berth.

Chapter 12

A Haunted House

In May of 1990, with Kim still trying to get pregnant, Sarkin decided it was time to go back to work. The thought had been growing for some time. He had been gone so long from his former life he believed work might reconnect him to the present. Happily, one thing his stroke had not disrupted was his chiropractic knowledge. Textbook learning and facts are largely undisturbed for most stroke patients, even as they struggle to focus on simple tasks and hold on to new information. In a way, Sarkin was better equipped to going back to being a chiropractor, than, say, doing the laundry. Kim, however, had reservations. After the stroke, she had hoped merely for Jon's survival, then she prayed that he might walk again, and finally that he might eventually work again, but she didn't think he was ready yet, either in mind or body. She worried about how he would interact with his patients. Would he know what to say? Could he concentrate long enough to listen to them? And finally, was he physically up to treating them? She argued that it was just too soon, that he didn't *have* to rush back. But she also knew that Jon *needed* to work. That much had been clear early in their marriage.

A few months after their wedding, they were driving to Logan Airport in Boston. Sarkin had been invited to give a professional talk, just before Halloween, and Kim was dropping him off. During the forty-five-minute drive she asked him when he thought about the future, what was the most important thing to him?

"Working," he said.

"Before family?"

"Yep."

"Before your wife and kids?"

"Yep."

At first she thought it was a strange answer and was somewhat put off, but then she realized that working was the thing Jon did best, and he had been happy doing it for her and Curtis, and for the children still to come. Her greatest fear in life was not having any children, which she no longer had to worry about. For Jon, it was not being able to work. No wonder he seemed so rootless and vague around the house.

In the spring of 1990, Sarkin resumed his practice, working just two days a week. The physical and cognitive strain of seeing patients, diagnosing their problems, and working out treatment plans exhausted him, and he needed an entire day just to sleep off the fatigue. At first, Kim drove him to the office, but he was determined to reclaim his independence and start driving again. His vision, hearing, and attention were all unreliable, but he persisted, and he finally passed the motor vehicle test on his third try. Still, the fact that he was alone in a car made Kim nervous, so much so that she refused to ride with him when he was driving.

Truthfully, Jon didn't feel comfortable driving, either, so he used the car only to go to and from work, and he always drove back roads to minimize the information his brain and eyes had to handle. Sometimes his friend Hank Turgeon would see him driving by, hunched over the wheel, with one eye closed, the better to cut down on the double vision. Sarkin's receptionist eventually offered to pick him up in the morning and drop him off at home in the evening, but when it became clear that his vision was not going to improve, Jon and Kim decided to move from Gloucester to Hamilton so he could walk to work.

Seven months after Jon returned to his practice, Kim announced she was finally pregnant again. They both hoped the birth of a second child would keep Jon moving ahead toward a full recovery, and also that another child might knit the family closer together again. But Jon was still a man going through the motions, each moment of his life an island of time where he was unable to feel the past or care much about the future. He wanted to show concern for his patients, wanted to be interested in hearing about Kim's day or what Curtis was learning, but his thoughts were usually drawn, instead, to whatever caught his eye at that particular moment.

Kim felt more and more on her own as well. Jon still had trouble being there emotionally for her, so she took classes in Transcendental Meditation, mainly as a way of dealing with the pain of childbirth, but also as a kind of therapy, a way to control her own stress and anxiety. When her water broke on August 18, 1991, she drove herself to the hospital. She'd already arranged for a friend, Kitty, to pick up Jon and Curtis and bring them as well. Kim decided early in her second pregnancy that Curtis should be part of the experience of welcoming a new member into the family, so along with the midwife, doctor, and nurses, Kitty held three-year-old Curtis during the delivery and Jon quietly stood by Kim's side. Robin Page Sarkin was born a few hours later, a happy, healthy baby girl. Months earlier Kim had refused her doctor when he offered to tell her the sex of the baby. She wanted to be surprised. Secretly, however, she had hoped for a girl.

Although the delivery had gone well, a half hour after Robin's birth, Kim began to hemorrhage. Normally after a woman gives birth, the placenta and fetal membranes slide out, allowing the uterus to contract and close off all the broken blood vessels. When the placenta fails to detach fully, however, the uterus can't respond and the blood vessels continue to bleed. In severe cases, like Kim's, unless the bleeding is stopped surgically, the patient can die.

Kim had hemorrhaged six weeks after Curtis's birth, which is why Robin had to be delivered in the hospital instead of at the birthing center next door. Now Kim needed an emergency dilation and curettage, which meant a doctor would have to scrape the remains of the placenta off her uterine wall. While the medical staff quickly prepared her for surgery, Kim seemed oblivious to the danger, still in a kind of postpartum high. She was confident the doctors would stop the bleeding.

"Is it really a girl?" she kept asking the nurses.

"You do know how serious the situation is, don't you?" one of them asked her incredulously.

Kim didn't really hear her.

"They've already been through a lot," an older nurse who knew Kim and Jon well scolded her assistant.

The surgery went well and in the months that followed, Kim cared for Robin just as she had for Curtis, rarely letting the baby leave her arms or her hip. She told Jon she understood better how hard it was for him to function with a physical disability, because with one arm always holding a baby, she essentially had just one arm to do everything else. She'd thought about Jon's difficulties when she was pregnant, too. The constant fatigue, the discomfort of even trying to bend over, all made her more sympathetic to Jon, and he liked how she shared these insights with him, and offered her advice on how to handle the exhaustion.

They had both hoped that the birth of their second child and Jon's return to work would help him continue with his recovery, but in the months that followed, the estrangement he felt from his former self continued. He felt so displaced that he had come to sense a kind of hollowness at his core, an unreality. Perhaps if he refamiliarized himself with past habits, he thought, he could find a way to slide back into his old skin. Kim supported him in the effort, but the more they tried to act like a normal family, the more apparent it was that Jon was profoundly changed by the stroke in ways they hadn't wanted to believe.

When the body is injured, there is usually some demarcation to the damage, an edge between what is healthy tissue and what is not. A wound to the sense of self, however, is harder to define. There is no visible scar in the brain, no gash in the terrain, just streams of electricity suddenly diverted or dammed up. The outer surface of the cerebrum may look untouched, but even the smallest of insults reverberates. The brain is like an Alexander Calder mobile; cut just a single string, and the whole structure collapses.

For Sarkin, time had become a blur. He was stuck living in an unending series of "nows," and whatever he was doing or feeling or thinking at that exact moment trumped everything else. Nothing flowed, nothing moved from cause to effect. The world had no priorities, and his attention knew no master. Rarely was he able to step back and see what had become of himself, but when he did, he felt something nag at him, something he couldn't quite define, like a person always searching for a word on the tip of his tongue but never finding it. He was a kind of shadow, a remnant of the former self he remembered but no longer recognized in the mirror.

Who he was now—or who he felt himself to be—was an amalgam, the parts hastily stitched back together but not quite fitting. When he'd awakened from his semicoma, all he knew was that he was alive. He'd had a kind of tunnel vision that protected him from realizing just how bad the devastation was. Now the picture of his life widened and the camera took in the whole scene. His view was that of a man returning home just after a tornado. The walls of the house remained, the concrete foundation was still there, but everything else was gone or askew, the bathtub in the backyard, the washing machine on the front lawn, clothes in the trees. Nothing was where it should be.

Freed unwittingly from time and place, Sarkin's mind became a kind of randomness generator. Words popped into his head, something he once heard, or read, a phrase like "low key." His mind would take it, soften it up, stretch it out, and turn it inside out. "Low key" became "Loki," god of discord and mischief in Norse mythology. Or he'd think of a line from a song but lose the thread and pick up another: "Mine eyes have seen the gory scenario of open revolt." Often, now, he wrote down what passed through his head, holding a kind of conversation with himself that had no beginning or end and was more a stream of consciousness of sound and words than complete sentences or ideas.

> Death is a thing that whistles away like a rusted freight train all covered with weedy overgrowth and if we are not paying attention it creeps up like locusts in a city and then the wind sweeps through the leaves reminding us of a thought we had but now gone into space and sandy time and a distant energy of unbelief and magic and the contrasting screams of mystery and the blue bleakness of mnemonic clarity and the translucent transcendence of confusion and hard stone onxylike of dim obfuscation and halving bliss and circumnavigational mantras and the fidelity of longed love and this heart blood wafting through summer central implicit in its integral strength like florid fetid fecundity and semiotic holographed destiny hypergraphic logarhythmic destitute deconstructed and even if I was to tell you how it feels to be on this raft you would be a pilgrim and then we would be sanguine.

This wholesale scrambling of his neural architecture left Sarkin in a constant state of restlessness, searching for something he was barely capable of articulating, but which sometimes expressed itself in the strangest of ways. Sitting at home one day his thoughts turned to a trip across the country he took one summer in college. His most vivid memories were of the flat ribbon of road unwinding through the Southwest, listening to the radio, and the strangely powerful shapes of cactuses blooming in the desert.

Suddenly, Sarkin stood up and went to the phone. He looked up the number of the local florist, and put in an order. Several days later, a truck pulled up to the driveway. Inside was $500 worth of cactuses. Kim was both horrified and impressed. Horrified, of course, that Jon had spent so much money on the plants, but impressed that he'd been able to follow through on something like ordering from a florist and then paying for the delivery. For Sarkin, the plants reminded him of his childhood and that Dr. Seuss book with the Stickle-bush trees. They also reminded him of persistence, because cactuses lived in the most desolate of places and yet survived.

On another occasion, before he gave up driving, Sarkin decided to go to the florist himself. After spending $100 on crocus, tulip, and daffodil bulbs, he returned home and dug a couple of dozen small, randomly placed holes in the steeply banked backyard.

"Jon, what are you doing?" Kim asked.

"Don't worry. It's a surprise. Just wait until spring."

In April, May, and June, the yard exploded with a new color nearly every day—sudden bursts of lavender, yellow, orange, and red shooting out of the ground like slow-motion fireworks.

Sarkin continued to get up in the morning, dress in his khakis, jacket, and tie, and walk to his office, but going through the motions of normalcy couldn't mask the fact that he wasn't normal. He still lacked the focus to manage any oversight of the children. When Kim found herself behind schedule and was in a hurry to pick Curtis up from kindergarten one day, she thought maybe it would be all right to leave the baby with Jon. She'd be gone for only ten minutes.

"Do you think you can manage Robin?" she asked him.

"Yeah, sure," said Jon.

"I'll lock all the doors. Just stay in the playroom with her, okay?"

"Of course," he answered, and he meant it, too, but like so much else in his life after the stroke, promises were an "of-the-moment" kind of thing.

He had every intention of watching the baby, until something else caught his attention and he left the playroom. Suddenly alone, Robin began to cry, then she found the one door out of the house with a flimsy lock. By the time Jon realized she was missing, she had toddled out into the middle of the street. When he got there, she was unhurt, but it had been a terrifyingly close call.

Two years after returning to his practice, Sarkin could still manage only two days of work a week, and those two days continued to be highly stressful. His physical problems, especially his double vision and lack of depth perception, made reading X-rays difficult. His responses were slowed, so interacting with patients had become a chore, and adjusting and realigning bones was hard. No matter how many or how few patients he saw on those two days each week, he returned home physically and psychologically exhausted.

The hardest part, though, was facing the fact that his interest had waned. He didn't know why, exactly, it's just that he often found himself taking longer and longer breaks between patients, and instead of reading charts, he would doodle on his letterhead paper, sometimes those cartoon faces, or spiky cactuses, or just random squiggles. The more he struggled to maintain his attention, the more discomfort he felt. He and Kim were both still counting on a full recovery, but the progress seemed to have slowed considerably, and he worried that this was as far as he was going to get.

Often when he became despondent, he talked to Kim. The strength of their relationship was not only that they loved each other, but that they also liked each other. They were, each of them, the other's best friend. Even though Kim knew Jon couldn't completely understand her frustrations with him, she also knew that he was willing to listen. The difference now, of course, was that she had to try harder to get his attention. He couldn't pick up on her moods and wasn't good about asking her how she was feeling, but when she told him she needed to talk, he would sit and listen, and in their shared but separate miseries, a new, different kind of intimacy took hold.

Over time, Kim saw that the work situation wasn't getting any better and they talked about what it would mean for him to sell his practice. He was torn about what to do. Occasionally, when he walked the streets around his home, he'd see someone and think, "His walk is all wrong," and then he'd flash back to the hundreds of reports he'd written as a chiropractor: "So and so's carriage is abnormally rigid and extremely guarded, an indication of considerable pain, which was readily apparent when he attempted to seat himself, which he did with great care and caution."

Being a chiropractor was who he was, and if he wasn't doing that, then who would he be? The prospect of a life without work was anathema. The only other activity he regularly engaged in—and enjoyed—was drawing, but he couldn't see how that could become a profession.

Then, in the summer of 1993, when the entire extended family was on vacation in Jamaica, Sarkin's sister, Jane, suggested he submit some of his doodles to the *New Yorker* magazine. As the features editor at *Vanity Fair*, Jane said she could at least connect him with someone in the "spot" illustration department, which oversees the small, unusual drawings that are scattered throughout each issue of the *New Yorker*. Sarkin wasn't sure if his sister was just trying to make him feel like he was doing something useful. He didn't exactly regard his doodles as art, but they were different, amusingly naïve, and so when he got back home he sent a dozen drawings off to the magazine, expecting to hear nothing back.

When he got a call from the *New Yorker* a couple of months later, he thought it was a joke. The magazine wanted to buy eight of his illustrations, for $125 apiece. The first drawing, a ghostlike cactus, appeared in the Halloween issue that year; two more small sketches were published in the spring of 1994. He still couldn't see drawing as a career, but the fact that someone outside his family actually liked his art, and even paid him for it, helped him realize that he couldn't continue forcing himself to go to work every day. In any case, it wasn't just a matter of not enjoying being a chiropractor. He knew he wasn't being as effective a doctor as he should be, and the last thing he wanted to do was fail his patients.

In December 1993, Sarkin finally decided to sell his practice. He was seeing so few patients by that time that it didn't take long to tell them all. Though on some level he couldn't believe he was giving it all up, on another he felt relieved. With nowhere to go during the week, he began visiting stores like the Dogtown Book Shop, which sold used books, or Mystery Train Records, which sold LPs and singles from the '50s, '60s, and '70s. He liked the smell, the feel of used things, the warped vinyl LPs, the coffee-stained pages, the inscriptions, "This book belongs to . . ." or "Happy Birthday, from Mom." Fingerprints of the past were all over them. Somebody had read this book, listened to that music, wore those clothes, or sat at that desk. The weight of identity, the feel of personal history, embedded in time—even when it was not his own—appealed to him. Sometimes he sat and read a chapter or two of a popular book on quantum physics, because the subject reminded him of his own feeling of indeterminacy. He was here, and he was not here. He liked the idea behind the Law of the Excluded Middle, a basic of Logic 101 courses: p and not-p. One or the other was true, but not both. That was impossible. Or was it? In his case, he "was" and yet he also "was not." The paradox reminded him of one of his favorite short stories by Franz Kafka, "The Hunter Gracchus." In it, the main character has fallen from a rocky precipice while hunting and tells the mayor of the town, the burgomaster, that while he knows he should be dead, he is also clearly alive.

"And have you no share in the world beyond?" asked the burgomaster wrinkling his brow.

The hunter answered, "I am always on the immense staircase leading up to it. I roam around on this infinitely wide flight of steps, sometimes up, sometimes down, sometimes to the right, sometimes to the left, always in motion . . . I am here. I don't know any more than that. There's nothing more I can do. My boat is without a helm—it journeys with the wind which blows in the deepest regions of death."

Without work, Sarkin needed to find new purchase in life, but his soul seemed to be wandering, like that of the hunter Gracchus. Kim was completely supportive of his decision to stop working, and she hoped that without the worry of his practice, his progress toward recovery might even pick up again. They had pushed ahead with the plan to keep building their family, and Kim found great satisfaction in being able to do so.

Their third child, Caroline Ruth Sarkin, was born in August 1994, and Kim once again drove to the hospital with Jon in the passenger seat. Curtis and Robin arrived later in the company of Kim's brother Lenny and sister-in-law Mary. The delivery, thankfully, took only an hour and there were no complications, as there had been after the births of Curtis and Robin. Kim felt the family was now complete, and raising her children gave her a strong sense of purpose in life.

For Jon, though, his purpose now was far less clear. Without the framework of his profession, he worried about being restless. He remembered the little piece of clear red plastic he found in the street one day near his Philadelphia apartment when he was in college. He'd put it in his pocket and when he got home he taped it onto a hole in the kitchen window screen, then wrote over the masking tape, "If you look through this on Tuesdays you will see the truth." He wished finding a way out of his current condition could be that easy.

Somewhere in his own wounded brain the secret to his new self resided, but he had no idea how, or even if, he would ever find it. "I feel like I'm a haunted house," he sometimes said to himself. "I feel at home, but never comfortable."

The mystery of how consciousness arises, how tissue, fluid, and electricity provide an enduring sense of self, is only beginning to be solved. Throughout the history of science, researchers have often relied on the absence of something, a deficit—a chromosome, a gene, an attribute—to understand the true nature of the thing missing. Jon Sarkin was his own psychology experiment. He was both scientist and subject, doctor and patient, and his peculiar condition of estrangement touched on some of the most fascinating research into the mystery of the brain's remarkable nature.

Chapter 13

THE GHOST AND THE ICE PICK

S arkin knew his mind had shifted in ways he did not yet fully understand, and while he couldn't say what the changes were exactly, he felt pieces of himself were out of place. He was both different and incomplete. Not a self, as much as the pentimento of a self—his life haunted by shadows and suggestions of a former life, like an old portrait hidden beneath the layers of a new painting. For Sarkin, the question was whether he could fully uncover that old portrait. If his soul or self was a product of his brain, would he even know *how* to reconstruct it from the damaged pieces of his new life?

When it came to making a clear connection between the brain and an individual's personality, it took scientists decades to accept the evidence from stories like that of Phineas Gage—men and women whose behavior, mood, and thinking had been altered by brain trauma, who were no longer the people they had been before their accidents. For more than three hundred years Descartes's dualism, the separation of mind and body as two distinct substances, had ruled the airy precincts of metaphysics. Finding that connection, that middle ground where mind met body, had been for the most part a quixotic journey.

That quest changed radically in 1949 when British philosopher Gilbert Ryle referred to the concept of a mind within a material body as "the ghost in the machine." Ryle thought Descartes's dualism—an invisible secret agent responsible for all our mental acts—was preposterous. There was no intangible self, no homunculus directing a person's inner life. If there was, then who—or what—was directing the inner life of the homunculus? To avoid an infinite regress, Ryle suggested a simpler hypothesis. There was no mind or consciousness that arose out of a brain. No mysterious, levitating spirit called a self or soul. The mental and the physical were simply the same thing. Ryle's single substance, or monistic, theory was an outright rejection of Cartesian dualism, and therefore eliminated the problem of the middle ground. There was no "mind" separate from the brain; no super-substance called a soul or self—just a brain, a machine built out of tissue and fluid. The mystical mind was out; the hardwired brain was in.

Ryle's philosophical theory had some science behind it. The view of a single substance had gained increasing credence due in part to a number of experimental brain procedures devised to alter the behavior of psychiatric patients. In 1890 German researcher Friederich Golz found that he could substantially reduce aggression in dogs by removing their temporal lobes. Inspired by Golz's results, Gottlieb Burckhardt, the superintendent of a Swiss insane asylum, tried a similar procedure two years later on a handful of severely schizophrenic patients. Although two people died and one later committed suicide, the calming effects on the patients who survived convinced him of its potential as a treatment for the mentally ill.

In the mid-1930s, Yale researchers replicated Golz's results with chimpanzees by removing parts of their frontal and prefrontal lobes. When the findings were presented at a conference in London, Portuguese psychiatrist Egas Moniz decided to pick up where Burckhardt left off, with humans. By severing the connections between the neocortex and the thalamus, the deep-brain structure that relays sensory information, Moniz believed he could assuage aggressive behavior in patients with particularly intractable psychoses such as paranoia and obsessive-compulsive disorder. The operation entailed inserting two wire knives in either side of the brain and then performing a sideways cutting motion. In 1936 Moniz heralded his successful "leucotomy," or "white-matter cutting," at conferences around the world. Although his surgical career was cut short three years later when he was shot in the spine and paralyzed by a former patient, Moniz spurred other doctors to investigate the radical treatment.

On September 14, 1936, at George Washington University Hospital in Washington, DC, Walter Freeman, assisted by James W. Watts, performed the first prefrontal lobotomy in the United States on a sixty-three-year-old Kansas housewife named Alice Hood Hammatt. For most of her adult life, Hammatt had suffered from insomnia, anxiety, and debilitating depression. After undergoing the lobotomy, however, she seemed calmer and more controlled. Although Hammatt died from pneumonia in 1941, her husband called the five years after her lobotomy the happiest of their married life. His wife had not disagreed. Months after the operation, she told Freeman she was no longer suicidal, worried much less, and could finally tolerate and even enjoy activities outside the house, such as visiting friends and going to movies.

Within a decade, Freeman pioneered the transorbital, or "ice pick," lobotomy. Using only electroshock to stun his patients into unconsciousness, and without administering an anesthetic, Freeman inserted a probe just under the patient's eyelid and above the tear duct and gently hammered the probe through the thin bone of the eye socket into the frontal lobe. After swishing the probe back and forth, severing the white-matter connections, Freeman repeated the procedure on the other lobe. He later wrote that he regarded these lobotomies as "mercy killing of the psyche" and well worth sacrificing part of a person's "driving force, creative spirit, or soul." Freeman's last lobotomy took place in 1967, when he accidentally severed a patient's cranial blood vessel, causing her to bleed to death. In the intervening thirty years, he performed about 3,500 of the approximately 40,000 lobotomies that took place in the United States.

With the development of the first psychotropic drugs in the 1950s, especially thorazine and lithium, the ability to chemically restrain violent or aggressive patients made psychosurgery obsolete. Experimentation with other approaches to brain control, however, continued.

On a bright spring afternoon in Córdoba, Spain, in 1963, Yale University professor Manuel Rodriguez Delgado, wearing a tie and sweater, dress pants and leather shoes, stood in a small fenced-in ring and stared down an angry, 550-pound bull. The animal, named Lucero, was just a few yards away and had been purposely worked into an agitated state by several professional bullfighters. Now, the forty-eight-year-old scientist was alone with the mammoth, angry animal. A home movie captured what happened next. Lucero charged at the intruder, but when the bull was a mere four feet from spearing the scientist with its horns, Delgado pointed at the animal, as if ordering it to halt, and at the same time pressed a button on a radio transmitter concealed in his other hand. Lucero skidded to a stop, then turned and calmly trotted away.

A few days earlier, with the permission of the bull's owner and breeder, Delgado had implanted a radio-controlled electrode just behind the animal's frontal lobe in a part of the structure called the caudate nucleus, which inhibits aggression. In the middle of the bull's charge, Delgado sent a surge of electricity into the animal's brain, stimulating the area that dampens aggression, thereby rendering the animal harmless. The experiment was not only a powerful and dramatic demonstration of brain manipulation, but also confirmation of the localization of personality characteristics—in this case aggression, which was tied to a tiny portion of the midbrain.

A few years after Delgado's unusual experiment, two Harvard researchers borrowed Delgado's radio stimulator and implanted the electrodes into the brain of a woman with severe epilepsy. Instead of placing the receiver into the caudate nucleus, they located it deeper in the limbic system, next to the amygdala, which, when stimulated, excites the neurons responsible for anger and aggression rather than inhibiting them. As their subject calmly played a guitar, the two researchers switched on the radio transmitter and she suddenly smashed the instrument against the wall.

Here was more hard evidence pointing to material monism—the theory that even the most evanescent of human traits, a momentary burst of anger, could be traced directly to brain matter. The theory that all things mental could be traced back to the physical flourished in the twentieth century as the brain gave way to further inspection, revealing billions of neurons and trillions of synapses. Like pulling threads, neuroscientists began to unwind the great skein of gray stuff where our deepest thoughts, desires, and memories reside. Magnetic resonance imaging allowed them to trace the origins of altruism, tolerance, and compassion to the prefrontal lobes. Fear was discovered entwined in the circuitry of the amygdala, where most of the signals travel outward to the higher regions of the neocortex and few run back down to the amygdala, which is why it is so hard to reason oneself out of feeling afraid.

Hunger, thirst, even infatuation were traced to the caudate nucleus; facial recognition to the fusiform gyrus in the temporal lobe, and *lying* about facial recognition was found to emanate from the frontal and parietal lobes. In 2001, Gabriel Kreiman, then at UCLA's Cognitive Neurophysiology Laboratory, discovered just how local perception can be. He recorded a single neuron firing in a subject when the person was shown a picture of the character Curly from the Three Stooges. Another neuron lit up only for images of Bill Clinton. While it was impossible to find the exact same neuron in other test subjects, single neuron firings were repeated with other patients, and always the activation occurred with a neuron in the medial temporal lobe.

Over the past decade, not only have single neurons been correlated to specific perceptions, but complex cognitive processes and abstract ideas have been traced to well-defined brain regions. Abstract thoughts, complex emotions, and beliefs, even specific types of humor have now been tied to particular areas of brain tissue. Visual puns cause shape-recognition neurons to light up; nonsense and surreal humor excites an area in the hippocampus that processes novel stimuli; and cartoons about people activate the social parts of the prefrontal lobes.

In 2008, Rebecca Saxe, a neuroscientist at the Massachusetts Institute of Technology, was able to identify the area of the brain, the juncture between the right temporal and parietal lobes, which processes a person's *thinking* about another person's *thoughts*. Nancy Kanwisher, a distinguished professor at MIT, called this "one of the most astonishing discoveries in the field of human cognitive neuroscience. We already knew that some parts of the brain are involved in specific aspects of perception and motor control, but many doubted that an abstract high-level cognitive process, such as understanding another person's thoughts, would be conducted in its own private patch of cortex."

Even moral conflict has been found to be brain-based. Several years ago, a group of neuroscientists and philosophers at Princeton University attempted to pinpoint the neural sources of moral decision making. To do this they asked a group of volunteers to consider the "Trolley Problem" posed by MIT ethicist Judith Jarvis Thomson in 1986. In the first part of the problem, the subjects had to decide what they would do if they were faced with a runaway trolley headed toward a group of five people. If they could switch the train onto a track where there was only one person, would they do it, knowing that the train would kill that one person?

The second part of the problem had a similar setup: a runaway trolley was headed toward five people. This time, however, the subjects had to imagine they were standing next to a large man on a footbridge over the tracks and the only way to stop the trolley from killing all five people was to push the large man onto the tracks. The situations were slightly different, but in both cases, one person's life is sacrificed for the sake of five, and yet normally the two parts of the trolley problem elicit two different answers. Most people are willing to switch the tracks away from the five toward the one, but are *not* willing to push the man onto the tracks. The lead scientist in the Princeton experiment, Joshua Greene, who is now at Harvard, studied his subjects using real-time fMRI (functional magnetic resonance imaging) technology, which highlights blood flow to the brain, and found that in both scenarios all the subjects showed increased activity in the anterior cingulate cortex, normally activated when a person has to respond to a conflict or dilemma. But secondary parts of the brain were also excited. When making the utilitarian decision about whether to switch the trolley from one track to the other, the volunteers showed increased activity in the dorsolateral prefrontal cortex, which is associated with cognitive control and abstract reasoning. When making the decision about whether or not to push the large man onto the tracks, however, very different brain areas were engaged: the medial prefrontal cortex and the posterior cingulate, both of them related to emotion and intuition. Two different kinds of moral judgment engaged two different parts of the brain. In deciding on a reasoned, utilitarian action for the greater good, the part of the brain involved in more controlled cognition was activated. In deciding in favor of a person's individual right to life, the more emotional, instinctive part of the brain was involved.

Cooperation, trust, guilt, shame, pride, embarrassment, shyness, and the sense of fairness—all have been mapped in the human brain. But as researchers have continued to fill in the landscape of higher consciousness, matching neurons with specific thoughts, feelings, and personal attributes, two pictures of consciousness have emerged. For most of the scientists, each new discovery adds color and definition to their understanding of the brain, but for many studying the source of consciousness, the discoveries have led to more questions, and in many cases a rejection of monism. If the ability to recognize the Three Stooges's Curly is located in one area of the brain, and all the attributes associated with that recognition—memories of his movies, a sense of humor, et cetera—are located elsewhere, then how do they come together to define more than just recognition, but the "experience" of Curly?

Monism, for many scientists and philosophers, had become a

victim of its own success. In assuming thoughts, perceptions, and feelings were nothing more than matter, materialists pushed neuroscience ahead until they were forced to wrestle again with the idea they'd tried so hard to reject: How and where do the fragments of perception come together to make a mental state, a "Curly state"? Ryle did not live long enough to address the confusion (he died in 1976), but his philosophical and scientific heirs now face the very real possibility that we are in fact more than the sum of our mechanical parts, more than our cells and synapses, more even than our neurochemistry.

Beginning in the 1940s, Wilder Penfield, an American-born Canadian neurosurgeon, began to cast doubt on the monist picture of consciousness. He showed, for instance, that in the case of memories, destroying the part of the brain where they seemed to reside did not do away with them. One time he operated on a severe epileptic who had arrived at the Royal Victoria Hospital in Montreal in the hope that a revolutionary surgery Penfield had been performing would cure her. The procedure involved exposing the brain, attempting to find the areas that stimulated the aura that often preceded seizures, and then removing that part of the patient's brain. Because the brain is insensate, the patient was under only a local anesthetic for the part of the operation that entailed cutting through the scalp and the skull. Once the brain was exposed, it was important for the patient to be awake and able to report to the surgeon when the probe elicited an aura, and therefore the possible site of a seizure. In the process of locating the center of the seizures, Penfield's probe elicited many other responses from his patients, including sensations, elaborate memories, and occasionally hallucinations.

As Penfield touched the probe to the frontal lobe of this particularly severe epileptic woman, she reported a detailed memory from her past she'd previously forgotten: She was sitting on the right-hand side of the backseat of a car stopped at a railroad crossing. The window was partially rolled down, and she could hear the warning bell ringing and then the clacking of the freight train's wheels as it thundered by. She could even count the boxcars. When the train finally passed, the car she was in drove into town where she suddenly experienced the familiar smell of coffee brewing in a popular local restaurant.

Other patients reported vivid fragments of memories about giving birth; listening to the laughter of relatives in a room next door; a mother calling to her son; a man with a pencil in his hand leaning against an office desk. One woman said she saw herself sitting in a familiar room listening to the sounds of the city traffic and her children playing outside. Penfield, who also elicited single sensations, such as the movement of a finger or a limb, recorded the location of each response and eventually produced a map of brain functions that became the standard guide for neuroscience researchers for many years.

But Penfield made another fascinating discovery. When he removed the parts of the brain that were the source of seizures, including spots where, when he had probed them, a patient relived a scene from the past, that memory was not lost. Patients could still summon the memory, even after that patch of cortex was removed. Just as curious was the fact that the patients were aware that they were in an operating room, even as they were doing something that seemed to Penfield much more than simple remembering, but rather a reexperiencing of moments from their past. The neurosurgeon called this a kind of "double consciousness," in which the patient was both subject and object of his own perception. "The vividness or wealth of detail and the sense of immediacy that goes with its evoked responses," he wrote, "serves to set them apart from the ordinary process of recollection which rarely displays such qualities." The only conclusion, for Penfield, was that there was some independent mind-action at work, which was separate and different from the working of the brain. There were two realities, in essence, he said, one concerning the processes of the physical brain and a nonphysical reality with which the brain interacted. The phenomenon of a double consciousness, coupled with the fact that memories could not be entirely localized, forced Penfield away from materialism and back to a kind of dualism:

In the end I conclude that there is no good evidence, in spite of new methods, such as the employment of stimulating electrodes, the study of conscious patients, and the analysis of epileptic attacks, that the brain alone can carry out the work that the mind does.

Sarkin's condition was in some ways like that of Penfield's patients. He had a kind of dual consciousness. No longer the man he'd been and yet unsure who he now was, he was both observer of himself and the object of his observation.

To traditional monists the idea of a double consciousness is unacceptable. Even to consider the question of two selves is regarded as heresy, a retreat to the discredited idea of a puppeteer in the theater of consciousness. And yet science has continued to uncover compelling evidence that a dual consciousness within our brains is indeed possible. If a memory cannot be located in a simple patch of cortex, then where is it?

Chapter 14

A DOUBLE CONSCIOUSNESS

As a medical procedure, Penfield's surgical treatment of epilepsy met with only partial success, but others took an interest in what the epileptic brain might reveal about brain function and ultimately that most elusive of phenomena, consciousness. Could a person really be of two minds? If double consciousness did exist, what could it tell us about the durability, even reliability, of the sense of self? Some of the most dramatic evidence for noncontinuity, for dual selves, has in fact come from a particularly drastic treatment for epilepsy: a corpus callosotomy in which the corpus callosum, the white-matter tissue connecting the brain's two hemispheres, is severed. By essentially cutting the brain in half, surgeons try to prevent epileptic seizures from spreading from one hemisphere to the other, thereby reducing the severity and, it is hoped, the number, of a patient's seizures.

Pioneered in 1940 by William van Wagenen, a Rochester neurosurgeon, the corpus callosotomy gained wider acceptance in the middle of the twentieth century when medications failed to alleviate disabling seizures in the most extreme cases of epilepsy. While a rarity, the corpus callosotomy is still used.

For Cal Tech's Roger Sperry and Michael Gazzaniga, these split-brain patients provided unusually persuasive evidence for a split consciousness. Perhaps the sense of self and its stalwart singularity, so often assumed, was not only malleable but corruptible. Could one person really be of two minds, have competing consciousnesses? Gazzaniga wanted to find out.

On one level, the question, perhaps even the answer, made perfect sense. After all, the human brain comes in a kind of double package, with a right hemisphere and a left, a left frontal lobe and a right, and so on. The only brain structure that does not come in a pair is the pineal gland, tucked deep in the middle of the brain, which controls our sleep/wake cycle. At the time of the Sperry-Gazzaniga investigation, the right and left hemispheres of the human brain were known to control opposite sides of the body, so that an object seen only through the left eye, for instance, is processed by the right hemisphere of the brain, and vice versa. But are the two hemispheres different in other ways?

In 1960, Sperry suggested to a young neurophysiologist, Joseph Bogen, that he team with Gazzaniga in Sperry's lab to study a patient after a corpus callosotomy. Bill Jenkins, a World War II vet who suffered debilitating epileptic seizures, agreed to take part in the research. In one experiment, performed after Jenkins's surgery, Gazzaniga and Bogen presented information in the form of colors, words, and pictures only to the right side of Jenkins's visual field. This meant the information provided by those colors, words, and forms would be processed in Jenkins's left hemisphere, where language and analysis were known to reside. Each time, Jenkins accurately reported exactly what he saw when information was presented to his left hemisphere, but when it was presented only to his nonverbal, intuitive right hemisphere, Jenkins was unable to report anything, almost as if he were blind.

While it appeared that the right brain didn't know what the left one was doing, Jenkins was nonetheless able to use his left hand, as instructed by the researchers, to tap a telegraph key each time something appeared in his left visual field. Again, because the left side of the body is controlled by the right, Jenkins's right nonlinguistic hemisphere allowed him to confirm what he saw, reflexively, through his left hand, but not consciously or verbally. If information depended on linguistic expression, it was lost in the "silent" right hemisphere. The left brain, the scientists concluded, processed and labeled the world in ways the right brain was unable to, and because Jenkins's hemispheres could not communicate with each other, when his right brain was exposed to information he could not articulate, he simply reported nothing.

Later, at the University of California–Davis, Gazzaniga continued his experiments with split-brain patients, essentially founding the new field of cognitive neuroscience. One very special subject provided Gazzaniga an unusual window into the possibility of truly split personalities. In an article published in 1978, Gazzaniga and his colleague Joseph LeDoux described their work with a fifteen-year-old corpus callosotomy patient, known only by his initials, P.S., in order to protect the boy's privacy. Because of an injury to his left hemisphere when he was a child, P.S. had developed an unusually powerful right hemisphere. Although unable to vocalize information presented only to his right hemisphere, P.S. was nonetheless capable of some language processing in his right brain—enough, in fact, that his right brain seemed to have a "mind" of its own. The scientists began by asking P.S. a series of questions, always substituting a key word or phrase in the question with the word "blank," but then showing the missing word or phrase to P.S.'s left visual field, which meant he could process the information only in his right hemisphere. Knowing that P.S. could not answer verbally, the scientists provided him with a ready alphabet in the form of Scrabble tiles. To the first question, "Who *blank*?," P.S. spelled out his name "P-A-U-L." For Gazzaniga and LeDoux it was like a voice breaking through from the dark side of the moon. They had communicated with a person's "silent" right hemisphere. A number of questions followed: Who was his girlfriend, name a favorite person, what hobby did he most enjoy? To which he answered

"Liz," "Henry Winkler," the actor who played Fonzie in the TV sitcom *Happy Days*, and "car" for hobby. They asked him to spell out where he was (V-E-R-M-O-N-T), the next day of the week (S-U-N-D-A-Y), and when they asked him about his mood, his right brain spelled out "G-O-O-D."

The most surprising response of P.S.'s right brain was his answer to the question about what sort of job he wanted when he grew up. The scientists had already asked his left brain, to which he'd written, as well as spoken, "Draftsman," but when the question was posed to his right brain, P.S. slowly spelled out, "A-U-T-O-M-O-B-I-L-E R-A-C-E (R)."

This wasn't a case of visual information unable to be articulated by the right hemisphere of the brain, this was a case where the right and left hemispheres in one brain appeared to disagree about what the person wanted to do with his life. P.S., it seemed, had two selves, or as Sperry later commented, "two separate realms of conscious awareness; two sensing, perceiving, thinking, and remembering systems." Gazzaniga went on to report other cases of divided consciousness.

In 2002, he wrote about watching a young boy, just a few days after the surgery that severed the hemispheres of his brain, try to pull his pants down with his right hand while his left hand pulled them up. Coined "alien hand syndrome" in 1972, the phenomenon of what scientists refer to, less colorfully, as intermanual conflict, is rare, but is seen not only in some split-brain patients, but also in those who have suffered disease, stroke, or other injury to their corpus callosum. Bogen reported on a patient named Rocky who took forever to dress because after one hand buttoned his shirt the other unbuttoned it. Another callosotomy patient had difficulty dressing in the morning because every time she reached into her closet with her right hand, her left hand would grab something else. Interestingly, the emotional right hemisphere, which controlled her left hand, often chose more colorful clothes than the woman intended to wear.

In 1991, Alan J. Parkin, at the University of Sussex in England, published a report about a woman who suffered damage to her corpus callosum from a ruptured aneurysm. After returning home from the hospital, she found her hands were frequently in opposition. When making an omelet, her left hand would throw an uncracked egg, a whole onion, even the saltshaker, into the frying pan. When packing for a trip her left hand often took out of her suitcase what the right hand put in. An eighty-one-year-old woman from Boston suffered a midline stroke in her brain. At Massachusetts General Hospital the medical staff observed scratches on the left side of her face, and she admitted that her left hand sometimes tried to strangle her. Neuroscientist Daniel Geschwind wrote about a callosotomy patient at UCLA Medical Center who told him that on several occasions his left hand, controlled by his right hemisphere, suddenly struck his wife, much to the horror of his verbally outraged left hemisphere.

Several years ago, Dr. Joseph Giacino, at the New Jersey Neuroscience Institute, interviewed a woman with alien hand syndrome. Using her right hand, the elderly stroke victim attempted to play a game of checkers with Giacino, but her left hand kept interfering, making a mess of the board. Her doctors also observed her trying to phone her husband, dialing the number with her right hand, only to have her left hand hang up.

Oppositional hemispheres, however, sometimes work together. The late British scientist Stuart J. Dimond described a split-brain patient who, in the late 1970s, reported she had overslept one morning—until her left hand slapped her awake. The notion of compromise in dual consciousness speaks to a fundamental brain process that began to take hold in Jon Sarkin's life after his recovery. The stroke and its aftermath had injured both his cerebral hemispheres. Damage to his right had affected his ability to move, see, and hear on his left side. The damage to his left hemisphere, however, was what set him out on his journey back to himself. The physical injuries and illnesses hadn't harmed his ability to speak, but they had fundamentally altered his brain's capacity to explain the changes in his life. What made Sarkin different from most brain-damaged patients is that he was conscious of the changes in himself; he remembered everything, and yet he felt distant from his own memories. This was a condition his brain simply couldn't accept. For stroke victims like Sarkin, the disconnect between injured parts of the brain also becomes a problem the brain must solve.

Often when stroke, disease, or trauma causes an identity disorder, the patient attempts to fill in the missing pieces. Time and again Gazzaniga has watched his split-brain subjects create explanations or confabulations with their left hemisphere for the mysterious actions of their right. In a now-classic experiment, Gazzaniga asked his teenage volunteer, P.S., to pick from an assortment of pictures the one most closely associated with the images presented exclusively to one hemisphere and then the other. For the image of a chicken claw, shown only to P.S.'s right visual field, he responded by choosing the picture of a chicken with his left hand. For the image of a snow scene viewed only by the left eye, P.S. chose the picture of a shovel with his right hand. Both hemispheres, working independently, correctly choose the corresponding pictures, but when Gazzaniga asked P.S. to explain his choices, he responded, "Oh, that's simple. The chicken claw goes with the chicken, and you need a shovel to clean out the chicken shed."

P.S.'s left hemisphere had no knowledge of the snow scene, which was viewed only by his right hemisphere. So when the researchers presented P.S. with his answers—the chicken and the shovel—his left hemisphere essentially improvised, fitting in the picture of the shovel with the only available information he had about the chicken. The story the left hemisphere invented makes sense only in relation to the left hemisphere's limited "experience." Gazzaniga realized that the left hemisphere is more than just the seat of analysis and language, but also the "interpreter" of experience. In other words, one of the primary jobs of the left side of the brain is to make sense of perception— or the lack of it. Psychologists use the term "confabulation" for this imaginative filling in of the gaps in memory or perception. In cases where information, in the form of memory or sensation, is lacking, the brain's left hemisphere is somehow compelled to create an explanation, a narrative, to solve the cerebral lacunae.

This puzzle-solving function is almost poignant, the brain's desperate attempt to make sense out of a troubling, inexplicable situation. Like a child whose mother has forgotten to pick him up at school, the brain invents reasons for finding itself all alone: she must have had an accident, or there was a traffic jam, or she's actually waiting for him right now, only in a different spot. When Gazzaniga presented one callosotomy patient's right hemisphere with the word "Walk," the patient walked, but when Gazzaniga asked him why he was doing that, the patient's left hemisphere quickly responded, "I wanted to get a Coke."

Neuroscientist V. S. Ramachandran believes that Gazzaniga's left-hemisphere "interpreter" is peculiarly active in one of the stranger maladies exhibited by certain brain-damaged patients. Ramachandran tells the story of a patient who, after waking from a coma following a head injury, proclaimed about his mother, standing next to his bed: "Doctor, this woman looks exactly like my mother, but she is an imposter." Later, however, when his mother phoned him, the man had no problem identifying the voice as that of his real mother.

When looking at another person's face, ordinarily the fusiform gyrus in the temporal lobe processes the visual information and passes it on to the amygdala, the emotional center of the brain. In the case of the man awaking from a coma, the connection in that visual identification stream between the temporal lobe and the amygdala was damaged. The face of the man's mother looked familiar to him, but there was no emotional connection. There was visual recognition, but no automatic emotional response. So in order for him to make sense of the missing information—the missing emotion—the man's left hemisphere resorted to a clever ruse, essentially manufacturing an explanation: the woman who looked so much like the man's mother is certainly familiar, but she must be an imposter.

The tenacity and resourcefulness of the left hemisphere is illustrated in a rare condition known as reduplicative paramnesia, in which a patient believes that one place, such as his house, has been moved to, or copied to, a second place, such as the hospital. Gazzaniga once observed a case of this at Memorial Sloan-Kettering Cancer Center in New York. The patient, a woman who lived in Freeport, Maine, was being seen by Gazzaniga for a cognitive deficit caused by a brain lesion. The woman suffered from hemi-neglect, or loss of awareness of sensory events related to one side of the body. In a few cases, hemi-neglect results in reduplicative paramnesia. Although Gazzaniga detailed all the reasons why she was now in a hospital in New York City, the woman insisted she was still back home in Maine. Finally, he posed a question that he thought would surely stump her:

"Well, if you are in your house in Freeport, then how can there be elevators just outside the door to this room?"

The woman barely blinked.

"Doctor, do you know how much it cost me to have those put in?"

Memory, sensation, perception: all are part of the story of the self. While the right hemisphere of the brain assembles the details of perception and experience, the left hemisphere tries to understand it all. Damage or delete any of those right-brain functions, and the left hemisphere will work overtime to fill in the blanks. If nature abhors a vacuum, as Aristotle once wrote, then the brain abhors a mystery. Presented with a puzzle, it will solve it in the easiest, most efficient way possible to keep the narrative going—even if it must resort to fiction. If Sarkin's past identity relied on his logical, responsible left hemisphere, then his new self would have to find a creative route back through the right hemisphere. Forced to find a new narrative to his life, his altered brain needed to reimagine its way home.

Chapter 15

The Accidental Artist

S arkin shuffled slowly across the boardwalk and down to the beach, tilting with each step as his cane sank into the heavy sand. Just in front of him, Kim pulled a plastic wagon piled high with beach toys and towels. Curtis and Robin walked beside her and Caroline, just a couple of months old at the end of the summer of 1994, slept in a sling across her breast. When the weather turned warm each year, the family often spent the day at one of the beaches along the North Shore, and the fine white sand of Crane's Beach was a favorite. Embraced by acres of dunes, the beach sits beneath a whale-shaped escarpment called Castle Hill and visitors must hike across a long boardwalk that reaches nearly to the sea.

After finding a good spot, Sarkin slumped down in his beach chair and turned his gaze to the crooked coastline meandering northward. Scooped out of the hard New England granite thousands of years ago, it was a beautiful but brutal landscape—beaten by the wind, squeezed and cracked by ice—and the houses scattered along the cliffs seemed to have an air of desperation about them, leaning this way and that, as if clutching for purchase on the rocky slopes. Sarkin tried to focus on the horizon, but his lack of depth perception caused water, sand, and sky to bleed together, making it difficult to figure out where one ended and another began.

By midafternoon, the beach was dotted with small, intimate groups. Children played at the edge of the water as low-slung waves tossed pebbles up the wet sand and chased seagulls into the air. A sand-sculpting competition was taking place and many people mingled among the temporary mermaids and dinosaurs. Sarkin's eyes hopscotched across the scene, trying to piece together the blurry puzzle of colors and shapes. Bodies twinned and overlapped, snaking across the beach. Bathing suits became a large patchwork quilt, with disembodied colors—blue, green, yellow, red—smeared across sand and sea. Sarkin dropped his hand from the armrest and picked up a flat granite stone. Suddenly the glacial veins of the stone seemed to liquefy in his hand, and the colors leapt up in an almost violent demand for attention. Green shouted at him, and he thought of the emerald green anoles of his childhood and the stare-downs he would have with their ancient, unblinking eyes. The spiny backs of the lizards metamorphosed into cactuses and suddenly he was back in 1975 driving along I-8 near Yuma, Arizona, where the giant saguaro cast man-sized shadows over the vast sands of the Sonoran desert. The radio was blaring "Tumbling Dice," by the Rolling Stones, and then the news came on. Someone was stealing saguaros and selling them on the black market.

One green thought led to another, and then another. His mind was racing, but it seemed to have a direction now, a motion of its own, and he felt compelled to follow. Holding the flat stone in his left hand, he scoured the sand around his chair with the right, feeling for something he could use as a drawing tool. Buried just beneath the surface his fingers found an old nail, and without hesitating, he began to scratch and scrape it across the stone. Back and forth, back and forth, he searched for a picture, not thinking about it, not knowing what he was doing, really, just letting himself follow this compulsion to draw. This wasn't doodling. This was a drive, a need to create that he couldn't really explain, not to Kim certainly, not even to himself. The stone had made him feel alive with possibility, and suddenly the aesthetic boundaries of the world began to realign.

He was reminded of the film *The Agony and the Ecstasy* about the life of Michelangelo, which he'd seen as a kid. He'd been so moved by the artist's intensity, his feverish desire to draw, paint, and sculpt. When Sarkin was a teenager he'd dabbled in art, of course, and then after the stroke, unable to do so many other things, he had delved back into drawing. But this urge he now felt wasn't cerebral and it wasn't casual; it was physical, visceral, and nearly involuntary. He didn't know where it would take him, and he didn't care, he just knew he couldn't stop even if he wanted to—and he didn't. When Sarkin finished filling both sides of the stone with layers of angled, crosshatched lines, a sense of satisfaction and calm washed over him.

Then he threw it into the sea.

The image, the drawing—whatever it was—wasn't the point, and this realization took Sarkin by surprise. As powerful as the act of drawing was, the product was unimportant. The activity, the process, being in the throes of creative compulsion, that's what it was all about. He was like one of Gazzaniga's split-brain patients forced to improvise in his search for meaning and context. In some way, it was as if his creative right hemisphere had taken up the slack of his damaged, analytic left. Unrestrained, Sarkin's right brain was now working overtime, trolling the unrelenting stream of his consciousness, trying to find order and pattern on its own. Destabilized by the stroke, his brain was searching for a way back to the world using the best tools still at its disposal: instinct and art. The process of drawing, every scratch, color, and line he made, was a way for Sarkin to fill in the gaps of his fractured experience—not with confabulated stories like a split-brain patient—with pictures.

Chapter 16

PROCEED WITH ABANDON

While Kim cared for the kids, Jon sat at the kitchen table or in the bedroom, the restless center of a storm of colored pencils, Sharpie pens, and pastel crayons. Drawing began to take over his life, and while Kim was unsure where it would lead, right now it offered Jon an outlet, something to do with his life, and a focus she hadn't had seen in him since before the stroke. She drove him to a nearby art store for supplies and bought him sketch pads for his birthday. The pictures poured from Sarkin's fingers, spilling out of some deep, unconscious place and memories that had shaped his earliest impressions: the cactuses he'd seen in the Dr. Seuss book from his childhood; the lizards he kept as a kid; the tailfins from his mother's old Cadillacs with their sleek, simple lines; and parts of favorite buildings from childhood, the shimmering steel web of cables and catenaries holding up the Brooklyn Bridge and the terraced arches atop the Chrysler Building. He was creating a kind of visual memory bank, and his canvas was anything he could find, from the blond wood of his cane to the cinder-block basement walls, from napkins, to diaper boxes, to the inside of a matchbook and the back of a paperback copy of Shakespeare's *Richard III*.

There was no preparation to his art, no thinking about subject matter, just reflex, immediacy, and then repetition, hour after hour and page after page. Usually he'd start with a colored pencil or pen and start something on one page—spirals or zigzagging lines, or rows of letters—then start something new on another page, and only after he'd begun ten or eleven or twelve drawings would he go back and, using a new color, add to each one. Or he might start by drawing a face with one eye, then draw it again, only this time with twelve eyes. Sometimes what he drew wasn't recognizable at all, just endless rows of tiny cross-hatched lines, creating a dense visual fugue. At the dinner table, in the bedroom, in the car, or at the beach, Sarkin was never without a backpack full of paper, pens, and pencils. If Robin or Curtis was wandering around looking for attention, he would hold the child in his lap and continue to sketch with a savage intensity. Kim's only request: no more drawing on walls and no Sharpie pens in the car. Her sense of smell was particularly acute and the pungent Sharpies in a closed space gave her a headache.

The drawings were still primitive, more like purges of his chaotic sensations than true art. One of his first completed projects was a colored-pencil sketch based on a favorite photograph of his father. Sarkin was a teenager at the time, vacationing with his family at Beach Haven on the Jersey shore and his father had just caught a fish. In the photo, Stanley Sarkin is smiling broadly, holding his fishing rod in one hand and his prize catch in the other. The memory was a warm one for Sarkin because it was during those vacations that his father seemed happiest, the most relaxed, and therefore easy for his kids to be around. No demands, no disappointments with Jon's grades, or the length of his hair, or his choice of music. When Jon, Richard, and Jane were growing up, their father was a strict disciplinarian, and if he didn't like the way Jon responded to him, he'd tell him.

"Who do you think you're talking to," he'd say, "one of your friends on the street?"

And if he was really angry, and Jon replied with, "Yes, Dad," his father spit back.

"Say yes *sir!*"

Only after his father's death did Sarkin appreciate the fact that his father was from a different generation, when parental authority was thought to be the only bulwark against a child's descent into drugs, crime, or a life of dissipation.

Usually when he finished a drawing Sarkin would toss it aside, but he held on to the fish picture and the next time he visited his mother back in New Jersey, he placed the drawing on his father's grave. He couldn't say why, exactly. Perhaps it was a way of honoring the good memories, or maybe it was a silent declaration to his father, who thought success depended entirely on making money, that his life was now guided by something else.

Now that they no longer needed to live near his office, Jon and Kim decided to move to a bigger house, perhaps one with a studio so that Jon could keep up with his drawing and painting. They looked all over the North Shore, but nothing was quite right, with enough bedrooms and yet also a small work space for Jon to do his art. Gradually, Kim started to think that maybe Jon shouldn't have a studio at home anyway, that it would be better for him to rent a space somewhere so that he would have to get out of the house every day. Jon's drawings and paintings were nothing more than an outlet, she realized, but it was an all-important one and having a studio to go to was about as close to having a job as he was ever going to get again. Rituals were a way of keeping order, too, and the habit of getting up each morning and leaving the house to do something and then returning in the evening was infinitely more organizing than spending listless hours at home.

Every day, Sarkin woke up unsure of what his eyes and ears told him. Sometimes he experienced the world synaesthetically, one sense expressing itself in the form of another, and the strange associations would end up as peculiar poems in his drawings and paintings.

I saw my own mother
In a garden of roses
I heard the sun thunder
And fill up the valley
I smelled her hair growing
In the mouth of the river
I tasted the fruit
Of the tree of confusion.

He had taken up the artist's life, but distinguishing imagination from actual perception was still difficult. The stroke had blended the two worlds. For a brief time he kept a journal on his night table to record his dreams—not so much for the stories, but for the sensations and emotions, the parts of his reality that had been so upended. Frequently the nocturnal scenes ended up in his drawings and paintings, a blending of unconscious self with the hallucinatory images of his conscious life:

"I may not remember my dreams," he said, "but my dreams remember me."

In the summer of 1995 Jon and Kim, Curtis, Robin, and Caroline moved into their new home in Rockport, just a couple of blocks from a bus line. A few months later Sarkin rented a large studio in an office building in downtown Gloucester, an eclectic town of working-class fishermen, old Yankee bluebloods, and bohemian artists. The oldest fishing port in America, Gloucester is also right next door to the oldest artists' colony in America, Rocky Neck. Poets, painters, musicians, and actors have all called the area home, at least during the summer—from T. S. Eliot to Whoopi Goldberg, Winslow Homer to Mark Rothko, Herb Pomeroy, the big-band trumpeter, to Willie Alexander, a latter-day member of the rock band the Velvet Underground.

But even as he immersed himself in his new life, he mourned his losses. He become moody and seemed to battle depression constantly. Kim knew what was going on. She and Jon hadn't spoken the words yet, but the day they both feared had arrived and they were having trouble admitting it to one another.

"I don't think I'm going to get any better than I am now," he said one day.

"I don't think you are, either," said Kim.

The realization was hard for both of them. Kim mourned, too, for all the things Jon would never be or do. When she saw other men playing with their kids, it broke her heart. Jon couldn't carry their children. He couldn't play Frisbee with Curtis or ride bikes with Robin. He would never be able to take them skiing, or hiking, or mountain climbing.

When he came home from his studio in the evening Sarkin sat on the floor, leaned against the couch, and proceeded to drink several bottles of beer before promptly falling asleep. Kim watched, hoping he would eventually shake free of the sadness. Finally, one night, she confronted him.

"You have to stop drinking," she said. "I watched my parents drink and I don't want my kids to grow up with a father who drinks."

Even now, Kim could always reason with Jon, and he would listen. Trying to hold back his tendency to depression, to harness his self-absorption—that was difficult—but he felt he could control his actions and quit drinking. No problem. And he did, that day.

Sarkin also decided to establish more discipline in his life by giving his days structure and some degree of predictability. After taking the bus into his studio every morning he stopped in at the Savory Skillet, usually for a blueberry muffin and coffee, then spent the rest of the day in his studio. He might draw for four hours at a time, or it might be for just five minutes, but when he took a break, usually it was to the same places—a visit to a neighbor or the nearby library, or just a stroll around town and down to the ocean. The order might be different, but the elements of his days were usually the same.

One of his frequent destinations was the Cape Ann museum, on the same block as his studio. Inside, he often lingered over the nineteenth-century maritime paintings of Fitz Hugh Lane. Sarkin liked the fact that Lane repeated just a few themes: fishing ships coming and going, at dawn and at sunset, in storm-tossed seas and becalmed.

"He draws the same stuff over and over," Sarkin said. "That's what you do."

Alone, in his studio, that's what *he* did, too. The basement room soon became a reflection of his mind. Instead of fighting the rambling thoughts and sensations that rolled through his head, Sarkin let them go, and so he found the mess all around him freeing. The studio usually smelled of Sharpie markers and mold, with a slight hint of Comet, which he used to wash his hands at the end of the day. Layers of paint, chalk, grime, and ground-down pencil shavings coated everything, even his telephone. When he took even a few steps anywhere in the studio, the cheap plastic pens strewn across the floor snapped and cracked like gunfire.

Sarkin stapled his canvases haphazardly to the walls and let paint bleed into the baseboards. He planned nothing. When he began a painting or a drawing, he had no idea what it would be, or where it would take him—and he didn't care. When his kids visited him in the studio and stepped on the drawings he threw on the floor, he never yelled. When he spilled coffee on his artwork, he didn't wipe it off. In fact, he liked the footprints and the coffee stains. The accidents of life contributed to his art, finished it, he thought. "Yes!" he sometimes said to himself after one of his kids left a big muddy print on the edge of a drawing. "Now it looks complete."

Usually he burned incense next to the large digital clock that reminded him when he needed to catch the bus back home. Random stacks of drawings, Styrofoam cups and plastic bottles, old magazines, used books, and an old milk crate turned the room into a lunatic's idea of an obstacle course. He loved the chaos, reveled in it, actually. The Romantic poet William Wordsworth once wrote that when the senses are overwhelmed it causes a "spontaneous overflow of powerful feelings," but only when those feelings are "recollected in tranquility" can the poet assemble the right words to construct a poem. For Sarkin, nothing was recollected in tranquility. Art took place immediately, in the here and now, and it took place unreflectively. "Art is to be. Not to be is not a good choice," he wrote on one of his drawings. And so art tumbled off the page and onto the walls, floor, and door of his studio, which he filled with rock lyrics and literary quotes:

In art and dream may you proceed with abandon. (Patti Smith)

We work in the dark—we do what we can—we give what we have. Our doubt is our passion and our passion our task. The rest is the madness of art. (Henry James)

Sarkin returned to some of the ideas of his youth, concepts that back in college were topics of midnight musings, but of no apparent consequence to his immediate life. Carlos Castaneda, in *Journey to Ixtlan*, which Sarkin read in the 1970s, had promoted the idea that while we live in the world for just a short time we should try to experience it directly, without preconceptions, letting the flow of sensations wash over us. This was never something Sarkin could, or necessarily wanted to, do before his stroke, but it was all he could do now. A piece of his philosophical past had become a part of the engine that drove his every action in the present.

Everything Sarkin saw, heard, read, or remembered ended up in his drawings in one form or another. The words he usually added were stream of consciousness, and they poured out onto paper and canvas almost as if he was reacquainting himself with the English language. Sometimes the text was a sliver of conversation he overheard, like the ramblings of a drunken diner he once met at the Savory Skillet. The man was trying to tell him that there was no free lunch, but instead said to him, "The free cheese is in the trap." Sarkin loved that, and incorporated the sentence into many of his drawings. Mostly, though, he added random snippets of songs or poems, parcels of words he'd read long ago and committed to memory, suddenly springing back into his consciousness. He repeated them constantly, finding special comfort in rhymes.

I got the blues so bad it hurts my feet to walk
I got the blues so bad it hurts my feet to walk

I got the blues so bad it hurts my tongue to talk.

Often, though, he simply created his own meandering, nonsensical sentences, punning and playing on words, just for the sheer fun—and sound—of it.

For who would bear the whips and corn-rowed hues of spaced-out
    firmamental blues?
Bushels of belief, gnashes of fire, a brick-a-brac heart attack.
The wanderlust of Homer's bust of dust to dust
Of Custer's last stand of the sea and land of a high school band

But sometimes the lines made exquisite, and heartbreaking, sense:

The pain of being an adult is dulled now, dulled by the dull rusty
Spike lodged in my brain, dulled by this one-too-many mornings
    scene,
Dulled by my dull thoughts. My future's doubled-back on my
    past. My
Now's been left for dead.

Sarkin admired the work of modern artist Robert Rauschenberg, especially his collages created from found objects and his sculptures, which, like much of Andy Warhol's art, were neo-Dadaist combinations of modern materials and popular culture. Sarkin also enjoyed simply writing the artist's name—in script, in one fluid motion, without ever having to pick up his pen to dot an "i" or cross a "t." The name of the artist Vermeer was special for the same reason, because writing it entailed never having to dip below the imaginary line on which he wrote: no *g*'s, no *i*'s, no *y*'s.

"I'm just going to write until it makes sense," Hunter S. Thompson once said, and that's just what Sarkin did. Each painting, each drawing, each poem was a reclamation. Word by word, and stroke by stroke, he was reassembling himself, first taking possession of his memories, then the images of his past, and now the words and sounds. "As I am, so I see," the essayist Ralph Waldo Emerson wrote, "use what language we will, we can never say anything but what we are . . . I know better than to claim any completeness. I am a fragment, and this is a fragment of me."

Chapter 17

BRIDGING THE IMAGINABLE

The train swayed drunkenly as it sped through the ground-down cities of Connecticut. Sitting by the window in a pair of old jeans and a faded T-shirt, Sarkin was on his way from the North Shore to visit his mother in New Jersey with a backpack full of art supplies and a few clothes thrown in. He was feeling good about himself. The experience with the *New Yorker* in 1993 had encouraged him to send his sketches elsewhere and the *New York Times Magazine* and the *Boston Globe Magazine* each bought a couple of his small drawings in 1995 and 1996, respectively. Doing art was what he did now; it defined him. Perhaps not in the same, clear-cut way being a chiropractor had, but nonetheless his life was now circumscribed by two places: his home and his studio.

An empty sketchbook lay invitingly on his lap. He picked out a colored marker and drew a circle on one page, then a square on another. Then he took out another colored pen and went back over the pages, adding new images—faces, spirals, eyes floating in the ether, and a few of his favorite words, *Rauschenberg* and *mesa*. This was his style, a kind of continuous flow from one page to the next, adding layers of lines and colors each time through.

"You must work for a newspaper," said a well-dressed woman sitting next to him.

"Why do you say that?"

"It just looks like you're working on a puzzle."

Sarkin liked that. Yes, he was working on a puzzle, he thought. The pieces of him had been scattered by his stroke, and his life, like his art, had become a process of reassembly. It was hard to feel so unfinished, so incomplete, and yet the ordering force of his art was powerful and profound. Without it, who was he? He looked out the smudged window of the Amtrak train and let the trees, buildings, and lonely homes blur by him.

"Everything you can imagine is real," Picasso once said. For Sarkin this was true, and what he imagined, he could not contain. There was an urge inside him now not just to create, but to communicate. He began to stuff his drawings into envelopes and send them through the mail—to friends, celebrities, people he had just met, people he wanted to meet, people he thought he might like and who might like him. He called these missives "boltflashes" and they usually arrived in the mail of the unexpected recipient in torrents: large envelopes festooned with drawings, writing, and cartoon figures. Sometimes he skipped the envelope altogether and slapped an address and stamps onto the back of a canvas and sent it naked through the mail.

After installing a computer in his studio, he often took breaks from drawing or painting to write letters, stories, and poems, and then e-mail them to friends, a dozen, two dozen times a day, and usually with similar themes.

The first one, a poem, would arrive:

It is night.
The sky is as black as a tornado cloud.
The only light in this god-forsaken place is an occasional oncoming
*   car.*

North and west lie the desert.
I look out the window
And all I see is gray desolation:
A bleak wasteland,
A black yawning maw of absolution.

Then a few minutes later, another message:

It was on the edge of Yuma when I saw the sign: MESA MOTEL ONLY 4 MILES I was tired, and Tony was too Not only physically and mentally, but emotionally, Spiritually, Somewhere on that atavistic borderline that ambivalently and Ambiguously, Vaguely, Delineates numbness and ache and hopelessness Was I beyond re-demption?

Screw this self-reflective navel-gazing bullshit I thought Fuck being tired We had to get to the San Diego airport by nine.

It was night—early morning, really—and the sky was as black as a dark tornado cloud.

Often it was the same story, over and over: Sarkin, along with an old friend, Tony Spirito, driving through the desert. The musings were poetic, but they could also be self-referential and mocking.

I roll down the window and the cool desert air slams against my face. Is "slam" too harsh of a word here? I suppose, but I learned all about hyperbole in Miss Larson's sixth grade English class, and I guess heavy-handedness is the sine qua non of hyperbole. What-ever. Ok—I roll down the window and the cool desert air gently wafts on my face.

And then he would begin it all again:

Road signs advertised motels or roadside attractions ("Mesa Motel: 4 miles"; "See Canyon Falls: next exit"). It was night—early morn-ing, really, and the sky was as black as . . . a simile fails me now, as

I have just arrived at my studio and am in no mood to conjure up comparisons like this. Suffice to say that it was dark, and the only lights on this deserted stretch were an occasional—very—oncoming car. We were on the outskirts of Yuma.

The e-mails were not exactly communiqués, but rather test balloons, a long, uninterrupted interior monologue, almost as if Sarkin was trying to unscramble the jumble of thoughts and sensations that dominated his daily life and to connect them somehow to the life he'd lost. The repeated images were like the tattered shards of snapshots from his past. His job now was to retake the pictures, snap new photos of his world and from as many different angles as he could, and in this way try to make sense out of the new picture of this life.

He thought about another of his favorite Dr. Seuss books, *On Beyond Zebra!*, in which a young boy, Conrad Cornelius O'Donald O'Dell, invents additional letters beyond Z, letters with names like Floob, Yuzz, and Zatz. At the end of the book, there is a single but unnamed letter, which is composed of all twenty-six letters in the Latin alphabet, and next to it is this invitation: "What do you think we should call this one, anyhow?" The question had sparked his imagination when he'd first read it as a boy, enough to inspire him to write an answer, which he sent to Seuss. Sarkin no longer remembered what he wrote, or if he heard back. But the name Conrad Cornelius O'Donald O'Dell he did remember. In some ways that's who he was now. He was inventing it all, making it up, just like one of those split-brain patients of Gazzaniga; he was searching for meaning. The images, the stories, repeated over and over in his drawings and writing, were rehearsals. Making sense out of the two halves of his life, the before and the after of his stroke, meant he would have to invent a new self, and to make it real he had to send it out into the world. The urge was relentless, insatiable, and yet it was also inchoate. Sarkin couldn't articulate this search yet, but he felt it in every brushstroke, in every penciled line and penned poem. If he could just keep telling these stories, he might stumble into himself.

Have I told you that THIS is how I maintain my sanity, By babbling these obsesso-compulso-manic-maniacal ramblings?

No matter. You see, I surf at the edge of entropy and compassion, of ecstasy and emptiness, of empathy and antipathy, of street noise and road signs that point out motels and roadside attractions: Mesa Motel Only 4 miles; Canyon falls next exit.

It was night—early morning, really—and the sky was as black as a dark tornado cloud. The only light was an occasional oncoming car.

To the north and west of Yuma lies the desert.

Suddenly, I felt quite tired, not only physically And mentally, but emotionally, spiritually. In this throe of—how should I put it?—existential funk?

Seething dread? Literal fear? Resigned

Desperation? I thought about what this place must have seemed like before the white man

Came: a bleak maw of ineffability.

I rolled down the window, and gazed at this once-pristine desert.

The coolness of the night air felt good.

Chapter 18

ART BOY

After Sarkin's art appeared in the *New York Times Magazine*, he received a phone call from Andrew Corsello, a writer at *GQ* magazine, asking to interview him. Sarkin was dumbfounded but excited, and invited Corsello up to his studio in Gloucester. Six months later, the January 1997 issue of *GQ* magazine included ten fashion trends for the spring, tips on hot stocks for the upcoming year, and a lengthy profile of Jon Sarkin titled "Metamorphosis." Both Jon and Kim thought Corsello had captured the sense of artistic urgency—and the mental chaos—in Jon's life perfectly.

Several weeks later, as he was sitting in his studio, Sarkin received another phone call. This time it was Hollywood. Someone from Tom Cruise's production company wanted to know if the studio could buy the movie rights to his life. Sarkin wasn't at all sure what that meant. In fact, he wasn't at all sure who—or what—Cruise/Wagner Productions was exactly. So he went to his fallback position, which was always "I need to talk to my wife." That night, after dinner, there was another phone call. This time it was Paula Wagner, the head of Cruise's film company and a friend of Jane Sarkin. She wanted Kim and Jon to know that she thought their life story was inspirational, that it would make a wonderful movie, and that she and Tom Cruise wanted to give them a substantial amount of money for the movie rights to his life. Sarkin's brother-in-law, Martin O'Connor, arranged for Creative Artists Agency to represent Jon in the negotiations, and a month later, Kim and Jon signed a contract that gave them about half a million dollars.

Not long afterward, screenwriter Billy Ray and a passel of production assistants spent a week in Rockport and Gloucester interviewing the Sarkins and doing local research. Before long, everyone from the local *Gloucester Times* to Hollywood's *Daily Variety* was calling. Sarkin loved the attention, fed off it in fact, and he began to think that "fate or karma or destiny" had a plan for him, including possible greatness. "I didn't ask for this 'mojo,'" he would say, referring to his creative compulsion. "This is what the starry cards have in mind for this here pilgrim. Fame, and possibly fortune, will follow this fame, like dream follows night."

To celebrate this new life, Sarkin took a couple of his friends out to dinner several times, and not just to any local dive, but L'Espalier in Back Bay, Boston, one of the city's most expensive restaurants. They ordered the twelve-course meal, drank $300 bottles of wine, and smoked $100 cigars. For the most part, Kim let Jon exult in his newfound fame—although she put an end to the $1,000 meals. He had a sense of purpose to his life now, and that was a good thing, but she also felt him slipping away, even retreating in some ways from the progress he had made back to his old life.

Jon had already put away his ties, button-down shirts, and wing-tip shoes. Now he wore dirty, paint-streaked jeans and sneakers and T-shirts whose random splatter designs reminded him of the work of abstract expressionist Jackson Pollock. One day Sarkin took his sweater off and impulsively painted the words "Art Boy" on it, then proudly wore it around town. When he took a break from drawing or painting, he often roamed the streets around his studio looking for pieces of trash to put into his paintings—a gum wrapper, a restaurant bill, even seagull feathers. Sarkin's friend Hank sometimes referred to him, affectionately, as the "Wildman of Gloucester," because he never knew what he would do next. And as Sarkin often reminded him, neither did he.

For the most part, his urge to draw and paint had been a lonely exercise in compulsion that neither Kim nor the rest of his family seemed to understand entirely. With most of his days now spent in and around his studio, though, Sarkin became increasingly plugged in to Gloucester's art scene. Outside of his wife and children, he found companionship mainly with other artists, like Chris Williams, whose giant metal sculptures appealed to Sarkin's sense of the absurd. Williams once explained his method of doing art as, "I get a lot of junk together and it speaks to me and says, 'I want to be a lizard.'" His front and back yards are littered with the tools of his art, including horseshoes, pitchforks, and tractor seats, and a seven-foot-tall sunflower made from rotary blades sprouts from the ground like some *Alice in Wonderland* dream.

Sarkin was drawn to musicians, too, even though he could no longer play the guitar. Willie "Loco" Alexander, a singer, keyboard player, and sometime drummer, lives just a few blocks from his studio and Sarkin often stopped by his house when he took a break from painting. In 1971, Alexander was briefly a member of the legendary rock band the Velvet Underground after cofounder Lou Reed departed for a solo career. Later, Alexander became one of the originators of the punk music scene in Boston. Today he lives and works in Gloucester, composes music, plays with friends, and occasionally tours. The ceiling and walls of his attic, where he built a music studio, are plastered with images from rock 'n' roll album covers and posters, and the flood of colors is hallucinatory.

Alexander's door is literally always open and occasionally when Sarkin ascended the rickety stairs to his friend's attic studio it was to perform spoken-word poems set to music, or he would just listen to

Alexander and other musicians play and sing. He felt comfortable among these men, sculptors, painters, and musicians whose lives were ordered around their art.

Vincent Ferrini was another one, a man dedicated to the craft of writing and virtually nothing else. A near-constant presence around Gloucester, he had been writing poetry since the 1940s. Alexander said of the unofficial bard of Gloucester that he "looked like Henry Miller and sounded like W. C. Fields." Sometimes they all ran into one another at Captain Carlo's fish market and restaurant. Located down on the waterfront, Captain Carlo's was a gathering place not only to eat and drink, but also to play or listen to music on summer evenings.

Gloucester, even more than Rockport where he and his family lived, had become home to Sarkin. Art was his ordering principle, too, the thing around which he structured his day, his hours, his minutes. When he came to the dinner table at night, his hands were caked in paint and all he wanted to do was talk about his artwork. If he wasn't drawing or painting, he was irritable and restless. To Kim, his behavior sometimes felt like a slap in the face, a rejection of all the progress he'd made to get better.

"Sometimes I think you're lazy," she would say to him. "Sometimes I think you're using the stroke as an excuse."

"Sometimes I *am* lazy," he would answer. "You say I should try harder to be a normal, happy guy again. I say I can't. Yes, you can. No, I can't. Well, I'm just not going to try anymore. End of conversation."

And for the time being, it was.

How could it be that this man, who had been so self-motivated, didn't want to work to change his behavior? The stroke had done many things to Jon, but for Kim the worst part was that it had unmoored her husband from the family man he'd once been. She felt starved for adult conversation. Jon was capable of rising to the occasion, sometimes buying his wife flowers, or playing Candy Land with his daughters, or free-association word games, or he'd let them direct his artwork.

"Draw a square doughnut!" one of them would shout.

"Draw a Martian playing the guitar!"

And he would.

Kim knew there would be no more big gains for Jon like they'd experienced in the early days of his recovery. There would never be a miraculous return to the man she'd married. But there were little things, his ability to be more attentive to his family, which she felt were not out of reach—if only Jon just cared a little more about reaching for them. She wanted him to be inspired by his wife and kids to fight for a few more successes, no matter how small. The hope—that was the hardest part for Kim to let go. For Sarkin, however, this was just the new order of things and he felt there was little he could do about it except embrace it.

"You don't find your way," he wrote on one of his paintings, "your way finds you."

Chapter 19

THE INNER SAVANT

When trauma, disease, or stroke impairs the brain, scores of skills and attributes can be compromised: the ability to speak, to write, to reason; coordination of limbs; recognition of objects or faces; the capacity to express emotion or to understand the emotions of others. Diminishment has always been the rule, not enhancement. The principle seems eminently logical: as brain cells die, so do functions, which is why everything slows with age.

Subtract large swaths of brain tissue, and you subtract a part of yourself, except in those rarest of cases when a deficit is somehow offset by a new skill that appears as if out of the blue. Sarkin's post-stroke artistry put him in this latter category, in an exclusive class of people called "acquired savants." History has recorded only a handful of these individuals who, despite, or perhaps because of, a disease or injury to the brain, suddenly are possessed of extraordinary abilities.

On August 17, 1979, ten-year-old Orlando Serrell was playing baseball in Newport News, Virginia. He had just hit a ground ball and was racing toward first base when an errant throw from one of the fielders hit him squarely in the left temple. Orlando fell to the ground, momentarily unconscious. Otherwise, he was unhurt and was quickly back on his feet and playing again. When he returned home later that day he said nothing about the incident to his parents and never saw a doctor, although for the next few days he was plagued with severe headaches. When the headaches receded, Orlando noticed he had a peculiar new ability. Although he was of average intelligence and had no special schooling or tutoring in memorization, he could spontaneously and correctly name the day of the week for any given date in the past or the future. He was also able to recall what he did, what he wore, and what the weather was like, for every day following the accident. Over the course of the next three decades, these strange new skills stayed with him, including an ability to recall the letters and numbers of every license plate of every car he had seen since August 17, 1979.

The exact reason for Orlando's newfound skills remains unknown, although clearly it had something to do with the blow to the head he suffered that day on the ball field. Some scientists who have studied him believe his memory for details is reliant on "eidetic imagery," a right-hemisphere function that aids the formation of visual memories. When Orlando was hit on the left side of his head, they argue, the visual memory center on the right side was stimulated into a near-constant state of activity as a kind of compensation.

The human brain is a biological marvel of checks and balances. Like Noah's ark, virtually everything inside it comes in twos, one on the left side, one on the right. The brain's balance also extends to its electrical circuits, the system of chemical signals used to convey messages between cells. Some cells (excitatory) are responsible for stimulating the brain, and others (inhibitory) for restraining it. Like an overloaded socket, the brain would burn itself out if all its circuits were always "on." In the same way that a person can't eat, talk, and drink at the same time, the brain can't perceive, calculate, and decide at the same time.

When the brain's checks and balances are upset, its center of gravity shifts, like a car whose wheels become misaligned. Normally when traveling down a straight road, a driver needs only to make the slight-

est of adjustments. If he takes his hands off the wheel, the car will continue to move in a straight line. When the alignment is off, however,
the car will veer hard to the right or left. The human brain is no different. Damage one hemisphere, and its balance is skewed, the skills of
one side overriding the deficits of the other. Rarely, however, do these
compensations amount to enhancement.

In 1980, a psychologist in Southern California reported the strange
case of a Mr. Z. Born in Mexico to well-to-do parents, Mr. Z. was nine
years old when several men broke into the family's home. The intrud-
ers murdered the father and shot Mr. Z. in the left temple. The bullet
entered at the hairline and tore through the language, hearing, and
motor centers of the left hemisphere before exiting at the back of Mr.
Z.'s skull. The wound left him mute, deaf, and paralyzed on the right
side, since the left hemisphere of the brain controls the right. Gradually, Mr. Z. regained his hearing and much of the use of his limbs,
although the left-hemisphere injury continued to cause general right-
sided weakness. Cognitively, most of the damage was irreversible. He
relearned how to speak, albeit in a halting manner, but he was unable
to write sentences unless he was copying them and was essentially incapable of abstract thinking such as telling time, estimating distances,
and doing simple arithmetic.

Nonetheless, as an adult Mr. Z. found employment as a factory
worker, a gardener, and then a ranch hand. When he moved to the
United States to live with a sister, he not only learned to ride a bicycle, he could strip it, modify it, and rebuild it, all without instruction.
Mr. Z., in fact, was incapable of following someone else's directions
and yet whenever and wherever he rode his bicycle—sometimes at
distances of up to ten miles from home—he remembered the name of
every street and never got lost.

Orlando Serrell, Mr. Z., and Jon Sarkin all suffered trauma to the
left, analytic hemisphere of their brains, and all experienced profound
changes in how they visually experienced the world. Instead of the
normally unremarkable balance of the five senses, their brains shifted
so that vision, and in Sarkin's case color as well, took on a larger role

in the processing of their everyday experiences. For Orlando, heightened vision coupled with memory appeared to enable him to remember the details of everything he observed in his life. For Mr. Z., the shift from left-hemisphere functioning to right-hemisphere enhanced his ability to instinctively navigate his environment and manipulate objects in space. Certainly for Sarkin, his sense of sight was altered, but why or how his brain seemed compelled to articulate his experiences as visual art—as paintings and drawings, on paper, canvas, walls, and rocks—remained elusive.

Sarkin didn't question his impulse to create, but neither did he understand it. A similar conundrum presented itself to John Langdon Down at the close of the nineteenth century. The British physician was the superintendent of Earlswood Asylum for Idiots in Surrey, England, and in 1887 he gave a talk to the medical society of London in which he described the condition of mental retardation that eventually bore his name (Down syndrome). In that same speech he also coined the term "idiot savant" (now savant syndrome), referring to a number of children at the asylum "who, while being feebleminded, exhibit special faculties which are capable of being cultivated to a very great extent." Down described ten of these patients for his audience. Among them was a boy who, in just one reading, memorized all six volumes of Edward Gibbon's *The History of the Decline and Fall of the Roman Empire*, though he understood not a word of it. Another patient had a vocabulary of fewer than one hundred words, but could play more than four thousand pieces of music on the piano. Jedediah Buxton, who could not write his own name, was a lightning-fast calculator. Once asked what would be the cost of shoeing a horse if the job required 140 nails and the first nail cost one farthing and then each subsequent nail doubled in price, his answer, formulated without use of pen and paper, was 36 digits long: 725,958,096,074,907,868,531,-656,993,638,851,106 pounds, 2 shillings, and 8 pence.

Numbers and calculations were not merely strange gifts; they were obsessions. When Buxton was being tested by members of the Royal Society in London, he was taken to see the famous actor David Garrick in a production of Shakespeare's *Richard III*. Apparently, Buxton paid little attention to Garrick or the drama of the play, instead counting the actors' words and steps and later reporting to his fellow theatergoers that 14,445 words had been spoken during the performance and a total of 5,202 steps taken by the actors.

Buxton also had an exquisite sense of space. He measured an entire county consisting of about 1,000 acres by simply walking it, then gave its area not only in acres, but in square inches, and then reduced the square inches to "square hairs-breadths," coming up with the figure of 48 hairs-breadths to each inch.

Down's most famous patient at Earlswood, though, was James Henry Pullen, who at the time of Down's address before the London Medical Society was fifty-one years old and had been living at the institution for thirty-six years. Pullen was born to parents who were first cousins, and was nearly deaf and dumb. He never learned to read or write, could say only the word *muvver* by the age of seven, and had limited social skills. His chief pleasure early in his life was watching the neighborhood boys sail tiny wood-and-paper ships in puddles in the streets of South London, and eventually he began to carve his own small ships out of pieces of firewood.

At the age of thirteen, James Henry was admitted to Essex Hall, a home for the developmentally disabled in Colchester, and two years later he entered Earlswood. Pullen eventually became a woodworking apprentice and learned to make furniture, at the same time continuing to draw and build model ships. Once, Down arranged for Pullen to visit a nearby naval station, and his subsequent drawings and models became exponentially more intricate and detailed. Pullen had great difficulty understanding instructions from others, and often flew into rages, so he spent most of his day in the workshop, fashioning the smooth curved planks of his models by steaming and bending the wood. His unusual creative prowess became so well known that even the Prince of Wales sent him pieces of ivory to carve. At night, alone in his room, he drew ships with colored chalk, and during the day he turned those drawings into three-dimensional models. At the age of thirty-five he began building a 10-foot reproduction of Britain's most famous iron steamship, the *Great Eastern*. Seven years later, it was finished. The ship included more than one million hand-carved wooden pins and 5,585 rivets. Pullen even fashioned more than a dozen lifeboats, and the entire upper deck could be hoisted up by pulleys to reveal fully furnished staterooms and lounges. Like Mr. Z. a century later, Pullen's brain damage was counterbalanced by a finely tuned mechanical ability, and like Sarkin, he was possessed of an uncanny drive to create.

In 1914, Alfred Tredgold, a consulting physician for England's unfortunately named National Association of the Feeble-Minded, wrote about Pullen: "His powers of observation, comparison, attention, memory, will and pertinacity are extraordinary; and yet he is obviously too childish, and at the same time too emotional, unstable and lacking in balance to make any headway, or even to hold his own, in the outside world."

When Pullen died in 1915 at the age of sixty-nine, the brain of the genius of Earlswood, as he was often called in the press, became the subject of scientific inquiry. His cerebrum was removed intact and sent to Maudsley Hospital in South London, where pathologist Frederich Sano performed a detailed postmortem, searching for clues to Pullen's artistic talent. Could there somehow be a relationship, Down and Sano wondered, between brain damage and creativity?

Among Sano's findings: The fissures and folds of the left hemisphere were longer than those on the right, showing a lack of general development in the left frontal lobe compared to the right. Areas of the frontal, central, and temporal lobes, also on the left side, were poorly articulated. However, the occipital lobes, the center of visual perception, were well developed and showed greater complexity than normal. Sano concluded that Pullen's substantial occipital lobes, in view of the underdevelopment of the rest of his brain, accounted for his unusual artistic skills.

In 2005 scientists at University College Dublin, in Ireland, reanalyzed Sano's autopsy findings and also discovered that Pullen's parietal lobes, which help maintain a representation of the body in space and objects in the outside world, also appeared particularly well balanced. Today, neurologists regard the parietal lobes as essential to manual dexterity. The Dublin researchers concluded that Pullen was, in all probability, an autistic savant. Because the white-matter connections between the parts of his brain were also underdeveloped, they believe that Pullen's experience of the world was largely governed by isolated, instinctive, right-hemisphere perceptual processes.

Wisconsin psychiatrist Darold Treffert has been studying savants for more than forty years. He believes their unusual skills are the result of damage not just to the parts of their brains that process their experiences, but to the parts responsible for remembering. Humans have two general types of memory: declarative, which involves learned facts, and procedural, or working, memory, which has to do with learning a skill like riding a bike or touch-typing. This latter category of memory is less verbal, more intuitive, and unconscious. For example, when someone hasn't skied in many years, or ridden a horse, we tell them, "It's just like riding a bicycle. Your body will remember." In other words, once learned, the skills involved in skiing or riding a horse or bicycling become deeply ingrained. Treffert believes that in acquired savants the damage to higher-level cognition results in compensatory enhancement of the deeper-lying, more primary, procedural memory processes. In the way that swimming or driving a car becomes unconscious and automatic for those who have acquired the skills, remembering street names or the weather for every day of the year becomes automatic for these brain-damaged savants.

Another possible theory involves event-related potentials, the measurement of the initial, lightning quick, preconscious activity that occurs at the very beginning of a mental process. Normally this low-level activity gives way to the higher-level executive functions used in abstract reasoning, judgments, planning, et cetera. Some scientists believe that in cases of autistic savantism, where a higher cognitive level is rarely reached, the autistic person's brain is perpetually stuck in low-level, instinctive-activity mode, where finely tuned technical capabilities also reside—capabilities like visualizing, calculating, and remembering details.

The implications of this theory are profound, because they suggest savantism isn't, in fact, an enhancement of brain processes, but rather a rearrangement. Freed from more complex executive-function tasks like abstract reasoning, the savant brain merely makes use of what is there and hyperfocuses on these low-level instinctive functions. If that's the case, then all of us are potential savants. Our brains have the capacity to hyperfocus as well, it's just that they suppress these functions, or balance them out, so that we can maintain normal lives. As complicated as the human brain is, in many ways it is quite simple: one thing at a time, please.

Chapter 20

BLINDSIGHT

After his stroke, Sarkin exhibited no peculiar ability to calculate dates or remember license plate numbers. His left hemisphere had been injured in numerous ways, but he clearly was not damaged intellectually. And yet some shift had taken place in his brain, something had been released, or was overcompensating for what had been lost, so that he was now refocused on one thing and one thing only: creating art. Drawing and painting became as instinctive and as necessary as building boats was for Pullen or taking apart and rebuilding bicycles for Mr. Z. The impulse was automatic and it was constant. Sarkin liked to say that he couldn't *not* do art. That was where his mind was every waking moment. He even claimed that was where his mind was in his dreams.

Although Sarkin's case was very different from Mr. Z.'s because his intellect remained intact after his stroke, Sarkin's brain was battered and bruised nonetheless, and when it finally healed, it was also deeply different. Enhanced, refocused, overcompensating—no one knew for sure what was going on, but it did seem to be in a constant state of alert.

"Brain plasticity" is the phrase scientists use today to describe the brain's capacity to change as a response to the environment. A few years ago, a British study revealed that a majority of London taxicab drivers have brains with expanded areas devoted to spatial navigation. The researchers hypothesized that years of negotiating the confusing landscape of the old city had probably helped to enlarge this part of the taxi drivers' brains. Gray matter increased in proportion to the need and use of one particular brain function—in the case of London taxi drivers, the ability to get from point A to point B.

The concept, if not the language, of brain plasticity was suggested in 1762 by the Swiss thinker Jean-Jacques Rousseau when he wrote that children's brains needed to be exercised as much as their bodies. Disillusioned with the economic and social inequalities of prerevolutionary war France, Rousseau rejected Descartes's mechanistic view of the world. Like the British philosopher John Locke before him, he argued that each of us is, at birth, a *tabula rasa,* a blank slate, which experience inexorably fills in. Human nature is not fixed, but rather acquired through learning and interaction with the environment. We are, in essence, works in progress. For this reason, Rousseau believed in both the malleability of children's minds and the perfectibility of man.

In 1783, five years after Rousseau's death, Italian anatomist Michele Malacarne attempted to prove the Swiss philosopher's hypothesis that brains could be "exercised," and therefore changed, by learning. Using pairs of birds born from the same clutch of eggs and pairs of dogs from the same litter, Malacarne intensively trained one half of each pair and gave no training to the other. After several years, he autopsied the animals' brains. Those that had undergone training had significantly more folds and fissures to their cortexes, which suggested to Malacarne that learning had changed their brains.

Whatever import Malacarne's research had at the time was lost, because for the next nearly two hundred years neuroscience was dominated by the belief that brains don't change in response to outside influences—they only shrink with age. In the 1970s, psychologist

Mark Rosenzweig proved otherwise. At the University of California–Berkeley he performed an experiment similar to Malacarne's two centuries earlier. For his study, Rosenzweig provided a group of rats with an interactive-rich environment and isolated a second group by placing them in small, empty cages. Those raised in the impoverished environment showed few changes in their brains. Those exposed to repeated stimulation, however, developed thicker cortexes and weightier brains. Nurture not only trumped nature, it changed it.

New brain-imaging technology has vastly expanded research into brain plasticity. In one recent experiment, German scientist Arne May divided twenty-four volunteers into two groups. A dozen of the subjects were taught to juggle over a period of three months, while the other twelve acted as controls and received no training. The brains of the volunteers were scanned both before the experiment and at the end of the three months. Those who had been taught how to juggle showed significant increases in white-matter connections between two areas of the brain, one involved in vision, the other in motor coordination. No changes were observed in the brains of those in the control group.

Since the discovery and acceptance of brain plasticity, researchers have noted many real-world examples of brain changes, but perhaps none more extraordinary than that of Ben Underwood. Movies taken of Ben as a teenager show him rollerblading down the street, shooting hoops, playing kickball, and chasing his siblings around the house in their Sacramento, California, home—all despite the fact that he was completely blind. In 1995, after Ben was diagnosed with a rare retinal cancer, both of his eyes were removed. Two years later, at the age of five, he was sitting in the backseat of the family car.

"Did you see that building?" he suddenly asked his mother.

"Yes," she said, but how did *he* know it was there?

Unbeknownst to his mother, Ben had developed highly sensitized hearing in the two years since his eyes had been removed. He "saw" the building because he was able to "hear" it by noticing a subtle change in sound waves caused by the family's car as it passed by an empty lot and then the high-rise. To "hear" objects around him when he was in motion, Ben instinctively began to click his tongue whenever he moved. By bouncing the sound of the clicks off surfaces—furniture, cars, people, curbs—he successfully navigated his way through rooms, down sidewalks, and on the basketball court. Metals produced soft echoes, wood dense ones, glass sharp echoes. Ben could distinguish the difference between a computer and a TV, a coffee table and a couch, and the loudness of the echo enabled him to judge the object's distance accurately. Sadly, in 2009, after the cancer returned, Ben died at the age of just sixteen.

The word *echolocation* wasn't coined until the discovery in 1938 that bats emit a high-frequency ultrasound that allows them not only to fly at night but also to identify prey as small as a mosquito. Dolphins, too, use echolocation, producing up to six hundred clicks every second. One of the first documented cases of human echolocation was the blind British adventurer James Holman who, in the early part of the nineteenth century, explored the world by tapping a stick or cane and bouncing the sounds off the objects around him. Holman claimed that he could also navigate on horseback, using the clattering of the horse's hooves to identify where he was and avoid obstacles.

Scientists now maintain that in cases like Ben's, where one sensory mode of perception has been eliminated, the brain simply recruits those "unused" visual neurons to enhance a complementary sensory mode. This is exactly what researchers at the Berenson-Allen Center for Noninvasive Brain Stimulation at Beth Israel Deaconess Medical Center in Boston proved in 2004 when they imaged the brain of blind painter Esref Armagan. The fifty-six-year-old Turkish artist has been without sight since birth, which meant he had never seen an apple or a fish or a tree, and yet for many years he has painted detailed realistic pictures, even portraits, using the sense of touch alone. For a strikingly accurate painting of Bill Clinton, Armagan referred only to an embossed photograph.

To understand how Armagan could achieve such realistic results, Dr. Alvaro Pascual-Leone, the head of the Berenson-Allen Center for

Noninvasive Brain Stimulation and a neurologist at Harvard Medical School, asked Armagan to sketch while undergoing an fMRI. "Esref's visual cortex lit up during the drawing tasks as if he were actually seeing," Pascual-Leone told the *Boston Globe* in January 2006. "His scan, to the untrained eye, might look like the brain of a sighted person." Had Armagan's brain grown new nerve connections, or had it simply enlisted neurons that were otherwise lying fallow?

To answer the question, Pascual-Leone and his colleagues constructed an experiment in which sighted subjects agreed to be blindfolded for five days and were taught Braille, as well as how to perform certain tasks only by touch. When their brains were scanned after just two days, small bright patches in the visual cortex lit up as they read Braille or performed a tactile task. After five days, the entire visual cortex was aglow. Yet within hours of having the blindfolds removed and repeating the same tasks, the volunteers showed no activity in the visual parts of their brains.

Pascual-Leone's conclusion was clear: This wasn't a case of new connections growing literally overnight, but rather the brain making use of neurons that weren't being stimulated by the sense of sight. Like a house in which some of the rooms have been closed off, the brain reclaims the furniture for use elsewhere. In this way, brain plasticity refers not only to the physical changes that take place with learning and new experiences, but also the brain's ability to rewire itself when damaged—to recruit healthy neurons nearby to perform the tasks that injured and dead neurons no longer can.

While the damage to Sarkin's left hemisphere did not affect his use of language or his ability to speak, it did affect how he used words. They tumbled from his mind into his art and his writing, where he explored their sounds, their textures, and whatever came next, but least of all their meaning.

Remembrances of things to come, when the stars are cool like falling water and I'm on the other side of the street, whistlin' down cabs and the muddled puddles splish-splashin' muddy dirty brown chocolate grit water on my feet and as they do, making this splatty sound and passerbys don't notice and the skies are gray with just a dull hope of blue—dusky gloom envelops the city and the rain turns to slush that will soon be snow. I know—that's backwards, but this is the way reality, so-called reality, is for me these days. Reality—that's a laugh. Reality sees ME, more than I see it. For me, reality is an unavenged angel, a hope of that pandora never dreamt about, not on her wildest night, in a fevered dream where the bed bugs are Kafkean. Did I ever tell you about the caftan I have that was made by Kafka?

Chapter 21

SEEING IS BELIEVING

How we make sense of the world starts with our eyes. There are five major areas in the brain devoted to different aspects of vision, including the size, shape, and movement of objects and the familiarity of faces. Our sense of touch uses just 8 percent of our cortex, hearing only 3 percent, but together the brain's visual systems make up a third of the brain and are responsible for interpreting about 40 percent of all incoming information.

In its human embryonic form, however, the brain begins as nothing more than a flat plate of cells that soon curls into a tube. During the first few weeks of gestation, this neural tube subdivides into seven or eight segments that make up the hindbrain. In one segment, nerve fibers develop to control the throat and tongue; in another, nerves form to move the muscles of the face, and in yet another, the nerves responsible for the coordinated movements of the eyes—the left looking left, and the right looking right—begin to grow. In the fully formed brain, the cells that control sight dominate. The retina is, after all, a part of the central nervous system. With 100 million photoreceptor cells, the retina compresses the lines, shapes, colors, and movement of the world and sends the information down one million fibers along each optic nerve, right to the thalamus at the core of the brain. As the brain's central relay station, the thalamus then exports the information from the optic nerves out to the various visual areas in other parts of the cortex. Human vision is so complex and so efficient that we need only 12 to 20 pixels in order to recognize a face, whereas a computer needs at least 150.

Yet what we know of the world through our eyes is hardly complete. Vision amounts to bits and pieces of light, a mathematical puzzle the brain must solve in less than the time it takes to blink. The retina is composed of seven tissue-thin layers of cells at the back of the eye where light signals are converted into neural signals. One retina can take in a million individual perceptions at once, discriminate among 8 million grades of colors, including 500 shades of gray, and perform 10 billion calculations a second—judging whether an object is near or far, moving or stationary, familiar or new. We don't see an object so much as make a very educated guess about it.

Your brain needs 30 milliseconds to detect the color of this book in your lap, 40 milliseconds to detect its edges, and at least 80 milliseconds before it can assemble the pieces—the colors, contours, edges, and surfaces—to figure out if the assembled picture matches anything familiar. With vision, nothing is whole. Everything is broken down, sent to different parts of the brain for analysis, and only then reassembled in the primary visual cortex—a half teaspoon of brain tissue that contains some 300 million neurons. "Resemblance is what is caught by the imagination as nature passes by," says neuroscientist Gazzaniga.

We see with our brains, not with our eyes.

This was confirmed several years ago by two neurobiologists at the University of Utah when they scanned the brains of students as they watched segments of the Clint Eastwood spaghetti Western *The Good, the Bad and the Ugly*. Their brains lit up like pinball machines, revealing that different parts of each student's cortex were watching "different" movies. When there was a close-up of Eastwood, for instance, an area called the fusiform gyrus, which is responsible for face recognition, was activated. When the movie was just scenery, the part of the brain that helps us navigate in three-dimensional space lit up. Of course, none of the students experienced the film piecemeal. Instead, in the split second it took for their brains to tease apart the images, their brains also put them back together again and continuity ruled.

"The mind is a kind of theatre, where several perceptions successively make their appearance; pass, re-pass, glide away, and mingle in an infinite variety of postures and sensations," the Scottish philosopher David Hume wrote in 1740. Hume's "perceptions" are what cognitive scientists today call "sensations," and the continuity of successive sensations constitutes the mind's understanding of reality. In truth, the way we normally experience the world—as a seamless whole of successive sensations—is only a trick of the brain. The magic of continuity is nothing more than a series of "stills," like thumbing through a flip book, where the ever-so-slightly different images on each page merge to create the feel of a moving picture. The brain does this to make sure our visual perceptions make sense, and in so doing, it is forced to leave things out. For instance, if you look up from this book and gaze across the room to a tree outside the window, you don't experience a blur of images as your eyes pass over every object between the chair and the tree, because your brain filters those images out. Likewise, when you blink you don't experience the 80 milliseconds of darkness when your eyes are closed. And yet, during the average eighteen hours we are awake every day—and blinking our eyes—we experience a total of nearly fifteen minutes of darkness.

Those university students in the movie experiment watched *The Good, the Bad and the Ugly* with their brains, not their eyes. A patch of brown became a horse, a black oval became a hat, and in a split second, an "Aha" moment in the brain rendered it all into Clint Eastwood on a horse. Yet the belief that we see the world as it is, unmediated and whole, persists—a necessary illusion created by our brains, which not only smooth out our jumbled sensations but also match them to our expectations. In a very real way, we see what our brains "want" us to see.

In an experiment of "intentional blindness," Daniel Simons and Christopher Chabris of the University of Illinois Visual Cognition Lab recently confirmed this "expectation" aspect of vision. The researchers asked volunteers to watch a video of a basketball game and to count the number of passes made by the team wearing white shirts. At one point in the video, a man in a gorilla suit slowly walks across the court while the game continues around him. After watching the video, the subjects were asked about the number of passes between the players, and most did quite well. Then they were asked about the man in the gorilla suit. Almost half said they had not seen him, even though he had blocked parts of the action.

The experiment not only showed that we can be "blind" to one thing when we are paying attention to another, but also that we tend to ignore things we don't expect to see. By habit and expectation, our brains are primed to interpret information according to a commonly accepted story line. This "confirmation bias" is what led the student subjects to discount the man in the gorilla suit—it simply didn't match up with their experience of basketball games.

Sarkin's brain no longer followed the usual visual rules for seamless perception. Instead of the usual expectations and confirmation biases, his brain saw everything at once and did little to prioritize the information. In the play of perceptions that passed through the theater of Sarkin's mind, the curtain was always up and the "characters" of sensation always onstage. No longer able to depend on habit or expectation, his brain tried to assemble new stories from fractured bits of his vision that didn't always make sense.

In some ways, Sarkin was the neurological embodiment of Claude Monet's impressionistic philosophy. The French painter spent his career in search of "the moment." Standing in front of Rouen Cathedral, he went through one canvas after another every hour, hoping to catch the light just before it shifted. Fourteen canvases in one day, ten at a time, forty in a single series: the cathedral at dawn, at noon, at dusk, bathed in sunshine, and enshrouded in mist.

"I wish I had been born blind and then suddenly gained my sight,"

Monet once said, "so that I could begin to paint without knowing what the objects were that I could see before me." Monet tried to paint the way Sarkin actually saw the world, as objects reduced to pieces. "Here an oblong of pink, here a streak of yellow, and paint it just as it looks to you, the exact color and shape, until it gives your own naïve impression of the scene before you," Monet once told one of his students. All of Sarkin's impressions were "naïve impressions." They were unfiltered and direct. Artists like Monet search their whole lives for this kind of immediacy, to capture an experience, a sensation, in all its color and light and texture, just as it is happening. Sarkin couldn't help but do the same thing. In fact, he was helpless to experience the world in any other way. There was an ever-changing present and nothing else. "The now is truly all I have," he sometimes said. "My stroke stopped the world and since my perception of *my* world is *the* world, the Now becomes eternal . . . This has everything to do with my art."

For this reason, Sarkin was attracted to the poetry of Wallace Stevens, because Stevens, like Monet, was constantly trying to express the inevitability of change and the impossibility of completeness, "of things that would never be quite expressed / Where you yourself were never quite yourself / And did not want nor have to be."

"Life should be about the experiencing of life, not the meaning," Sarkin said, but that meant accepting incompleteness and indefiniteness, and he wasn't sure yet he could. He wanted to know who he was, *what* he was, but he suspected Stevens knew better.

"Both in nature and metaphor, identity is the vanishing-point of resemblance," the poet wrote. "The prodigy of nature is not the identity but the resemblance and its universe of reproduction is not an assembly line but an incessant creation."

Chapter 22

SENSES AND SENSIBILITIES

Psychologist and philosopher Nicholas Humphrey believes vision is more than just a cerebral construction, an amalgam of expectations. Vision, he claims, is active and creative and therefore fundamental to the understanding of consciousness. Humphrey uses a single simple example: the color red. We don't see red, he says, we "do red." When we look at a blotch of red light on an otherwise blank movie screen, we are generating a particular state of consciousness. We are having a red sensation or, as Humphrey describes it, we are "redding." The sensation didn't exist before we looked at the red light on the screen, and the sensation will disappear as soon as we close our eyes or walk out of the room.

The primacy of sensation in our cognition, says Humphrey, can be traced back to our evolutionary ancestors who were essentially reactive creatures, using their ears, nose, eyes, skin, and tongue as tools of survival. If they saw a predator, they hid. If they felt cold, they huddled for warmth. Even today, some of our sensations elicit vestigial responses. When we experience a sudden fright, we instinctively draw our shoulders up. The quickest way for an animal to kill a large prey—or an ancient human ancestor—is to sever the carotid artery in the neck so that the victim bleeds out quickly. When we hunch our shoulders, we're forming a natural barricade against attack to our carotid arteries.

Imagine living off the land and foraging for edible plants. In order to tell the difference between two similar-looking flowers, say Queen Anne's lace and the water hemlock, we might look at their stems. Queen Anne's lace has green stems. Those of the water hemlock are streaked or spotted with purple. We could compare their height (hemlock is taller), look for the small purple floret in the middle of the hemlock's umbrella-shaped bloom, or even smell them. Hemlock carries a distinctively musty odor, Queen Anne's lace the odor of carrot greens.

Thousands of years ago, sensations played a crucial role in human survival. If a piece of meat smelled rotten, best not to eat it. Today, sensations still play that role—throw out the sour milk, avoid a hot stove—but they also guide everything from our appreciation of fine art to our choice of coffee. Sensations are no longer strictly tools of survival, a means to an end, demanding an active response. Instead of requiring action, says Humphrey, sensations are internal, so when we are looking at that red blotch, the crux of what it is like to be you or me, at that particular moment, in that particular place, is "redding." Sensation is at the core of who we are. "What matters is that I feel myself alive now, living in the present moment," Humphrey writes. "What matters is at this moment I'm aware of sounds arriving at my ears, sight at my eyes, sensations at my skin. They're defining what it's like to be me."

Contrary to Descartes, Humphrey claims consciousness is not grounded in thought, but in bodily sensation. The relationship between sensation and consciousness was something Humphrey first started thinking about back in the 1960s when he was a PhD student in physiology and psychology at Cambridge University in England. His supervisor was an American psychologist, Larry Weiskrantz, who at the time was trying to confirm that the destruction of the visual cortex produces total blindness. To do this, Weiskrantz created lesions in the visual cortices of monkeys. One day he went off to a conference and Humphrey, ever curious, decided to spend time with the monkeys to see if they were really as blind as Weiskrantz and the others in his lab thought they were. What Humphrey had noticed was that occasionally one of the blind monkeys appeared to reach for a piece of food as if it could actually see the food. So he played with the primates, waved pieces of food in front of them to see if they would grab the food, and noticed that when he moved his hand back and forth in front of their faces, their eyes followed. With just a few days of training, he got the monkeys to take a piece of apple from his hand, something they could not do when the lights were turned off in the lab. On Weiskrantz's return, Humphrey announced that he had taught the blind monkeys to see.

Weiskrantz was impressed and the two continued the research together. For the next seven years, Humphrey spent much of his time, both inside and outside the lab, with one particular blind monkey named Helen. He played games with her and took her for long walks in a nearby park. Eventually Helen was able to run around a room, all the while avoiding obstacles. She could successfully forage for fruit and nuts on the floor. She could even catch flies in midair.

Helen had no visual cortex, but she had something *like* visual perception. Weiskrantz called it "blindsight," and in the 1970s, he conducted a series of visual experiments on a blind human patient in his thirties, referred to in the published research as D.B. At the age of twenty-six, D.B. developed a brain tumor that necessitated the removal of his right occipital lobe, essentially blinding him in his left field of vision. The right hemisphere of the brain not only controls the movements of the left side of the body, but vision in the left eye as well, and vice versa. Despite the removal of half his primary visual cortex, D.B. was able to correctly point to objects in his "blind" field and describe their shape and orientation, even though he maintained he could not "see" them. Instead, he claimed all he was doing was guessing.

Around the same time as Weiskrantz was doing experiments with D.B., Humphrey met H.D., a twenty-seven-year-old woman, blind since the age of three, who had recently undergone cataract surgery. Afterward, her doctor assured her she was now capable of seeing normally again, but she continued to believe she was blind. Unlike D.B., H.D. had not had a substantial portion of her visual cortex removed, but the primary vision-processing area of her brain had been dormant for twenty-four years. Like a limb that withers without use, her visual cortex had degenerated. For different reasons, H.D. was in the same boat as D.B. Both believed they were blind because their brains were either unable, or incapable, of interpreting what they saw. Their visual sensations were like islands of information cut off from the rest of the world.

Eventually, Humphrey decided to try to teach H.D. to "see" the same way he had taught the monkey Helen. They took walks together all over London, with Humphrey holding her hand and describing the sights. Over time, he says, she began to point out pigeons and flowers and successfully cross the street on her own. But to Humphrey's astonishment, the experience of seeing, for H.D., was decidedly disappointing. Somehow it lacked depth and body, and was not at all the way she had imagined. She told Humphrey she "simply could not feel it." She could sense things around her, she acknowledged, but because her brain still had trouble putting those things in context, the world seemed vaguely unreal. H.D. eventually gave up on the experiment, put her dark glasses back on, and returned to being "blind." If sensation was the critical element to consciousness, Humphrey concluded, then lacking full-bodied visual sensation, H.D. lacked an important part of herself.

In 1972, Humphrey published an article about Helen the monkey and the phenomenon of blindsight in *New Scientist* magazine. The cover story bore the headline "A Blind Monkey Who Sees Everything." Humphrey's own title of his article, "Seeing and Nothingness," was a play on the famous philosophical text "Being and Nothingness," by

French existentialist Jean-Paul Sartre, and in many ways, Humphrey's article was deeply philosophical. To be, to have a self, to be a first-person subject, according to Humphrey, means you must be the subject "of something," you must be engaged in perceiving and sensing. None of us lives in a vacuum. In the same way that seeing red means one is in a state of "redding," being a self implies action, and for Humphrey that action is clearly sensation. "What sensation does is to track the subject's *personal interaction* with the external world—creating the sense each person has of being present and engaged, lending a hereness, a newness, a me-ness to the experience of the present moment."

Humphrey argues that sensations not only were the primary tools of living for ancient man, but as life got more complex, it also became important to remember these sensations, to have inner representations of them, so that primitive man could understand what was happening to himself and also to others. In determining who and what was friend or foe, we needed to remember and predict, and so we moved from a present-tense world into one with a past and a future. *I remember how that striped animal attacked my family a while ago; I better watch out for it again.* The more we advanced evolutionarily, says Humphrey, the more sensation became less a tool for survival and more of a private experience, a part of consciousness.

Sensations in this context are hardly objective, measurable phenomena. Instead, because of their interior nature, they are susceptible to the vagaries of experience: My redding is not your redding. The role of experience in sensation was one of the central issues of the eighteenth-century Enlightenment. On a Saturday in early July 1788, Irish scientist William Molyneux had just finished reading an abstract of the philosopher John Locke's just-published *Essay Concerning Human Understanding*. Inspired by the British empiricist's argument that knowledge is acquired through experience, Molyneux wrote to Locke asking him to consider a thought experiment: Imagine a blind man regains his vision and two metal objects are placed before him on a table. One is a cube and the other a sphere. Previously, the blind man knew the difference between these objects by touch alone. Could he now distinguish them by sight alone? Could his brain "translate" the knowledge from one sense, that is, touch, to another, vision? The question, really, was trying to get at the crux of whether sensation informs perception.

There is no record that Locke ever answered Molyneux, but more than four years later, after the two men had become acquaintances, the Irishman again posed the thought experiment to the famous philosopher. This time Locke answered and published their exchange in the second edition to *Essay Concerning Human Understanding*:

"I agree with this thinking Gent . . . and am of opinion, that the Blind Man, at first sight, would not be able with certainty to say, which was the Globe, which the Cube, whilst he only saw them: though he could unerringly name them by his touch . . ."

Both Molyneux's question and Locke's answer were entirely hypothetical, but in 1728, forty years after Molyneux first queried Locke, William Cheselden, an English physician and anatomist, published the first account of a blind person having his vision restored. Despite the removal of the patient's cataracts, wrote Cheselden, he could neither understand the difference between the shapes of things nor could he distinguish one thing from the other by sight alone. Sensation, how he experienced the shapes of those things, powerfully influenced his perception and his ability to distinguish between objects.

More than 270 years later, Mike May, like H.D. and Cheselden's patient, had his vision restored. At the age of three, he had lost his sight in a freak accident when a jar of fuel exploded in his face. His left eye was so damaged it was removed and his right was useless from intense corneal scarring. But in 1999 a California ophthalmologist offered a way back to sight: stem cells and a corneal transplant. On May 7, 2000, the bandages were removed from May's eyes. Fewer than a half dozen people who have had their vision restored have been blind since childhood. May was part of an even rarer group—he now had near-perfect sight through the new cornea in his right eye. But his visual cortex, unused to seeing for so many years, was creaky and sluggish. In the first few months after the successful surgery, May's inability to distinguish objects and faces was more proof of the validity of Locke's response to Molyneux.

Locke believed that only through raw experience did our sensations have meaning. Today, however, scientists know that visual perceptions are dependent not only on the environment, but on the neurons of the primary visual cortex where visual information is processed. Because May did not use his visual cortex at an early age, it did not fully form, and because it did not fully form, he lacked the complete panoply of visual tools the average person uses every day to process the information coming through the retinas. Today May still can't distinguish the faces of his wife or two sons from those of strangers. Even his own facial expressions are limited, because he has never been able to view those of others adequately. For this reason, he has difficulty understanding the moods, personalities, and emotions of other people, too.

When May was blind, few people would have said he wasn't a fully functioning person despite his handicap, and yet oddly the restoration of his sight highlighted what was missing in his relationship to others. He did not feel the absence of these things, because he could not know what he could not sense. And yet it would be hard not to think of May as somehow incomplete. Sensation isn't sufficient for personal identity, but it is necessary. If our ability to sense the world is compromised, so is our sense of self. None of us lives in total isolation. Even if we were stripped of all but one sense, that single avenue into the world would provide the action and meaning to life, driving both memory and expectation and coupling one moment to the next, past to present, and present to future. Sensations build narrative. Experiencing the world through our senses, Humphrey says, not only allows us to feel alive and engaged in the most meaningful way possible, but helps each of us generate the story of our lives. Disrupt sensation, and identity is dislocated, too. Sarkin experienced these distortions every day. He did not see, hear, or even walk in the world the same way he had before his stroke. Instead, he felt imbalanced, incomplete, and restless. Making art was his primary mode of sensation now and his most instinctive, most intimate way of interacting with the world. To feel wholly connected to life again, he would have to reimagine it one sensation at a time.

Chapter 23

THE IMAGINATION INSTINCT

No one knows for sure why the human species is the only one to engage in art and storytelling. What do they accomplish? What *did* they accomplish? Forty thousand years after we left Africa, most of us still have a primal fear of snakes, which goes back to our ancestors' deadly encounters with them in the bush. We feel goose bumps when we're afraid, which thousands of year ago made our fur stand up. The effect was to produce an extra layer of cushion for protection and possibly the appearance that we were larger, and therefore more intimidating, to the potential predator.

But writing and art?

The biologist E. O. Wilson once asked whether artistic activities were "adaptations that directly improved our survival and reproduction? And if so, what exactly were the advantages conferred?" Harvard psychologist Steven Pinker takes the contrarian position, calling music "auditory cheesecake." Along with the rest of the arts, music is nothing more than a side effect of evolution, he claims, a kind of accidental dessert. Some evolutionary psychologists suggest otherwise. They believe that rather than being a by-product of evolution, the creative imagination has been an important adaptive tool in human survival.

The first known art appeared in abundance in human cultures about 30,000 to 40,000 years ago, around the same time one of the genes that codes for brain size appeared. In other words, our brains got bigger just as our artistry began to flourish. John Tooby and Leda Cosmides, directors of the Center for Evolutionary Psychology at the University of California–Santa Barbara, believe art developed from play and evolved in humans to help us learn how to survive. The prevailing view of evolutionary psychologists like Tooby and Cosmides is that we are neither blank states waiting to be filled up by experience nor are we completely in the thrall of our genetic makeup.

We come into the world equipped with the primitive software necessary for emotions, perception, and cognition, and our experiences expand that software through learning. But if we had to experience everything personally in order to learn from it, few of us, say these evolutionary psychologists, would make it into adulthood. Instead, every piece of art, whether painted or written or sung, is a story, a kind of virtual reality that we not only can "try on," but by imaginatively "practicing," can use to greatly shorten an otherwise dangerously long learning curve.

Cave paintings of buffalo, for instance, perhaps wired the brains of our hominid ancestors to know which animals provided the best nourishment, how the members of a tribe needed to cooperate during a hunt, or how to kill the animals. If true, then art wasn't a by-product of human evolution, but a necessary, even pragmatic, aid to survival. Artists, as representatives of all human seekers, are in part compelled to use their imagination and create art as a way of finding or imposing meaning on experience. "There was not enough time for human heredity to cope with the vastness of new contingent possibilities revealed by high intelligence," writes E. O. Wilson. "The arts filled the gap."

If, as Picasso said, art is a lie that helps us see the truth, then our brains make artists of us all. In the same way a baby babbles in order to develop its capacity to understand and use language, we learn how to live by experimenting with life's possibilities through the world of our own imagination. Perhaps not coincidentally, children begin to play imaginary games at eighteen months, around the same time they develop a "theory of mind," the ability to understand and imagine the mental states of others. In the early years of human evolution, so much of survival depended on social awareness: who was trustworthy and who wasn't, who could protect you and who couldn't. Among animals with lengthy childhoods, especially birds and humans, play is universal, and it is compulsive. The most popular kinds of play for both animals and humans are pretend fighting and chasing and hide and seek. The individual and social benefit of practicing and learning both skills is evident: the survival of the individual, and by extension, the group. Play, and by substitution art, gives the audience or reader a high-density dose of social information that cannot be gained from single experiences in real life. In this new model of human cognitive evolution, the arts are a rehearsal for reality, increasing the odds of survival.

Our sense of our own selves is profoundly connected to our capacity to imagine—to the stories we tell, the songs we sing, and the pictures we paint. If there was a way back to himself for Sarkin, to identity, to his soul, he now had the right tools.

Chapter 24

GHOSTS OF THE PRESENT

Sarkin liked to say the "nidus" of his art was his stroke. He'd stumbled across the word in a medical textbook as he browsed the shelves in Gloucester's used book store. *Nidus* is the Latin word for "nest" and refers to the breeding place where bacteria, parasites, and other agents of disease lodge and grow. The idea of a breeding place, even for germs, appealed to Sarkin because in some ways that was still how he thought about his art, or at least his compulsion to make it—it *was* a kind of pathology, since it took hold of his life only after an accident, an insult to his brain. But like that little bit of irritating sand inside an oyster that produces a pearl, a nidus could produce something beautiful, too.

When he left home each morning for his studio, Sarkin felt the strange, inescapable pull of the sea. The water seemed to hover over the town, squeezed between the old clapboard buildings at the end of the street. "Like a tidal wave," he would say to himself as he teetered toward the bus stop. "Like a wall of water." And as he moved from home to studio, he inevitably felt the weight of one self slide away as he slipped on another. Most of the time he still felt uncomfortable in his own skin, caught between being a husband and father and being an artist, never completely one or the other. Making art was a powerful impulse—it was all he wanted to do, all he could do—but why he was doing it even he didn't really know.

Every once in a while, in the middle of the day, Sarkin left his art behind and walked to Gloucester's esplanade, where an eight-foot-high statue of a fisherman at the wheel looks out across the ocean. Sarkin would gaze at the thousands of names on the ten bronze plaques, a memorial to Gloucester's lost fishermen, and then read the inscription on the monument, taken from the 107th Psalm: "They that go down to the sea in ships."

He thought about all that had happened in his life, and a feeling of inescapability, of inevitability, sometimes washed over him. What happens over the course of a life is unpredictable. We spend our time unwinding the skein of experience without knowing what is to come, only knowing that it will. "Something about all these dead people," he said. And it was true, the ghosts of Gloucester were all about him. Not just centuries of fishermen, but thousands of artists and writers who were drawn to these rocky ledges by their harsh, unpredictable beauty. For Sarkin, the hardness of the landscape represented all that was inevitable about his life—the stroke *and* the art—and the unforseen yet constant effects of fate. He thought of Eliot's poem about this place:

> What might have been and what has been / Point to one end, which
> is always present.

Tethered to the present, Sarkin's sense of time was exquisite and acute: moment to moment to moment. He was conscious all the time that nothing stays the same, even when we're standing still. The moon tugs at the seas and the planet slows. Time stretches out. Tomorrow will be longer than yesterday. Charles Darwin was reminded of the inescapable effects of time 175 years ago as he circumnavigated the globe. In February of 1835, the twenty-six-year-old naturalist was more than halfway through his five-year expedition aboard the H.M.S. *Beagle* when the ship docked at Valdivia, Chile, on the western coast of South America. Walking through the coastal forest, he lay down for a rest in the shade, and as he drowsed, the ground began to rumble. A few seconds later, an earthquake rocked Darwin awake, and for two minutes the undulating earth made him feel dizzy. In his diary, he wrote:

> I can compare it to skating on very thin ice or to the motion of a ship in a little cross ripple . . . An earthquake like this at once destroys the oldest associations; the world, the very emblem of all that is solid, moves beneath our feet like a crust over a fluid; one second of time conveys to the mind a strange idea of insecurity, which hours of reflection would never create. In the forest, a breeze moved the trees, I felt the earth tremble, but saw no consequence from it.

Darwin rushed back to the ship, which eventually sailed 200 miles north to Concepción, where the damage to the city was considerable. All the while Darwin observed and recorded what he saw in the natural world around him: The land had been lifted several feet in Concepción and left beds of mussels decaying 10 feet above the new high-tide mark. Later he found seashells 13,000 feet up in the Andes mountains. All of it brought to mind the works of geologist Charles Lyell, which he had just been reading. It seemed to Lyell, and now to Darwin, that it could take millions of years for geologic processes to change the planet along with the planet's animals and plants. Not one earthquake, but millions. Not one death, but billions. And if it took that long to shape and reshape the geography of the land, then why not the forms of life that inhabit it as well—forms of life that would have to learn to adapt to slowly evolving environments. Changes in the Earth, particularly in animal life, were neither sudden nor dramatic. They were the offspring of grindingly slow forces that fashioned new species from individuals just different enough to be able to survive the capriciousness of nature. Death was, is, evolution's best friend, its driving force. Only the extinction of common ancestors allowed the few who were different, hardier, to survive. Natural selection is as much about weeding out unproductive or ineffective genes as it is about passing on those that aid survivability. If a mutation aids the health of the species, then it remains; if it does not, natural selection removes it from the gene pool.

But if death was the defining catalyst for evolutionary change, then why, as a species, did we develop the capacity to contemplate death, especially our own? How is introspection or self-awareness (the basis for consciousness) an aid to the human species?

Nicholas Humphrey argues that self-awareness evolved as a social tool, a way to understand the thoughts, behaviors, and motivations of others, so crucial for survival. Evolution simply favored those who could put themselves in another's position. By feeling angry, I am better equipped to understand and predict anger in someone else. If I can understand another person's motivations, then I can predict their behavior, and if I can predict their behavior, then I can also try to manipulate that behavior to my advantage. Survival thus favored those who were self-aware and introspective, who had developed consciousness.

V. S. Ramachandran also believes self-awareness and consciousness of others co-evolved. I see you in pain and I feel pain, and then I create a pain model in my mind of what it feels like to be in pain. In this way, consciousness of others is a self-reinforcing act, which also means that being cut off from others is self-alienating. For Sarkin, whose sense of self was so severely stressed, there was not only personal dislocation, but social. How could he experience the reality of others when he barely felt real himself? If consciousness of others is critical to consciousness of self, then Sarkin faced double the trouble.

Chapter 25

THE EXTENDED SELF

Rarely did Sarkin return home from his studio clean. The art was primal and raw and physical, and no matter how many canvases he painted or drawing pads he filled, his brain wanted more. What was it trying to tell him? In some ways, his frenzy for art was like the alarm that goes off when a person is in pain. Primates sense the intensity and the location of pain in the somatosensory cortex and the insula, but they experience its unpleasantness in the dorsal anterior cingulate. In fact, patients who have had their dorsal anterior cingulate surgically removed are able to report how intense a pain is but say they are no longer bothered by it. In these cases, it's easy to see the importance of this second pain center in survival. If our brain doesn't tell us it hurts, then we're not going to take our hand away from the fire or stop the kitchen knife from accidentally slicing into our thumb.

What is happening in the brain when we hurt ourselves is twofold. First, there is a sensation that registers in the somatosensory cortex, which then sets off an alarm—the "ouch," if you will—in the dorsal anterior cingulate. Here's where it gets really interesting. With the dorsal anterior cingulate screaming for attention, the right ventro-lateral prefrontal cortex comes riding to the rescue. This part of the brain is responsible for nonautomatic, conscious thinking and is especially active when someone is reflecting on himself—like when he's in extreme pain. The feedback mechanism comes into play because in order for the right ventrolateral prefrontal cortex (the RVPFC) to figure out how to respond to the stimulus—take that hand away from the fire!—it has to turn down the volume on the alarm in the dorsal anterior cingulate. This is why it's hard to think when someone is shouting in your ear. In the case of physical pain, the shouting is really coming from your own brain. The RVPFC is critical in figuring out how to respond to adverse stimuli. When it jumps into action, its job is to identify, interpret, and reflect upon the pain signal coming from the dorsal anterior cingulate. In cases of extreme panic and danger, we become immobilized when the prefrontal cortex can't respond because the alarm from the cingulate is just too loud.

This division of labor in the pain response means that strange things can happen if the connection between the cingulate and the RVPFC is disrupted. In general, the right ventrolateral prefrontal cortex is active when people reflect on their own personal traits or remember personally relevant information. The area is so crucial to self-awareness that in controlled studies when researchers disabled the right dorsolateral prefrontal cortex of volunteers with a brief, noninvasive electrical stimulation, the subjects had trouble recognizing their own faces. Increased activation of this right-hemisphere area has also been tied to negative emotional judgments in people with major depression.

Because Sarkin's left hemisphere had been damaged by his stroke, his right hemisphere had become more active, including areas involved in visual creativity. But it is just as likely that his right ventrolateral prefrontal cortex was overactive as well. If Sarkin's impulse to create arose out of an imbalance in his brain, then so, too, did the content of his creativity—the cactus, the Cadillac tailfins, even the plays on words, were all attempts at personal interpretation, as if making art was a response to some emergency that wasn't really there. There was no external crisis. Instead, Sarkin was stuck in that feedback loop between two areas of his brain, trying to make sense of his own mental and spiritual dislocation.

The same kind of disrupted feedback loop involved in the sensation of pain has also been implicated in the placebo effect, which happens when someone experiences relief from pain or an improvement in their health from a pill or treatment they believe to be therapeutic, though it is actually harmless and ineffective. Boston anesthesiologist Henry Knowles Beecher was one of the first physicians to observe and then study the placebo reaction. During World War II, in January 1944, Beecher landed with the Fifth U.S. Army on the beaches of a small Italian hamlet named Anzio, where the Roman emperor Nero built a villa in the first century. For four months, 35,000 Allied troops struggled to maintain the beachhead under withering German assaults. Medical supplies, especially morphine, were severely depleted and Beecher decided, on more than one occasion, to give wounded soldiers injections of a "placebo," a simple saline solution, as a way to at least calm those calling out for relief. The solution was nothing more than saltwater, of course, but to Beecher's astonishment, the injured men, who thought the syringes were full of morphine, felt substantially free of pain after receiving the injections.

After the war and back at Massachusetts General Hospital and the Harvard Medical School, Beecher conducted studies of the placebo reaction he observed on the battlefield, and became the first to urge that clinical trials make use of placebos to test the effectiveness of a drug by controlling for those subjects who improve solely because they believe they will. Placebo-controlled studies are now the gold standard for clinical trials.

In the late 1990s, in the heyday of new mood-controlling medications such as Prozac, William Potter, a psychiatrist at Eli Lilly pharmaceutical company, became concerned about a strange and inexplicable trend in the development of psychotropic medications. Potter was in charge of the early trials of new mood stabilizers, but of ten antidepressant trials, the drugs were unable to outperform the placebo in seven of them.

Potter's findings spurred new research into the placebo effect at the beginning of the twenty-first century. During the past decade, numerous research studies have revealed three medical conditions that consistently respond to placebo: Parkinson's disease, depression, and pain. In 2002, Canadian scientists found that Parkinson's patients release dopamine in their brains in response to placebo. At UCLA, scientists employed quantitative electroencephalography to image the brains of test subjects, all of whom were told they were receiving the antidepressant, though only half were receiving the real pill. The results indicated the same activation in the mood-controlling prefrontal cortex of the subjects whether they had taken an antidepressant or a placebo. In experiments involving heat or mild electrical shocks, scientists tracked the release of pain-dampening opioids in subjects all of whom believed they were being given an analgesic, though none actually was. In every case the expectation of relief was enough to create the same physiological response as drugs known to activate the analgesic centers of the brain.

The placebo works because the brain has separate areas for the sensation of pain and the expectation of its unpleasantness. Which means the mere anticipation of relief can result in the unpleasantness melting away. But that also means the brain can act in reverse, that is, an anticipation of a harmful effect can *cause* an unpleasant sensation or pain. Although it has long been the subject of anecdotal stories, this opposite of the placebo response, called the nocebo effect, has been verified.

One of the most famous cases involved a small-town physician by the name of Drayton Doherty. In 1938, the doctor admitted a sixty-year-old farmhand to a fifteen-bed, dilapidated hospital for blacks in Selma, Alabama. The farmhand had lost some fifty pounds and was unable to keep food down without vomiting. Tests for tuberculosis and cancer were negative, and despite a feeding tube, the patient's condition continued to deteriorate until he was too weak to talk and only semiconscious. It was then that the man's wife confided in Doherty that one night four months before, her husband had an argument with a witch doctor. At the time, many of the poor in rural Alabama believed in voodoo. The witch doctor, the wife explained, had put a hex on her husband and told him he was going to die and there was nothing any medical doctor could do.

The following day, with the patient's wife and relatives gathered around the hospital bed, Doherty told an entirely fictional story. He announced that he had lured the witch doctor back to the cemetery the night before and had choked him until he revealed how he had put the curse on Doherty's patient. The witch doctor, he said, told him he had rubbed lizard eggs into the black man's stomach and all but one died after hatching. One lone lizard, however, was now eating away at the patient, and because he had met the witch doctor, Doherty said he now knew how to cure him. With much fanfare, he injected a full syringe of an emetic into the arm of the patient, who promptly began vomiting. Minutes later, as the man continued to retch, and when no one was looking, Doherty surreptitiously slipped a real lizard into the basin of vomit and showed it to his patient.

"Look what's come out! You're cured!" the doctor exclaimed.

Surprised and speechless, the man fell into a deep sleep and when he awoke twelve hours later, ate heartily. Within a week he was discharged, and after several more months, was completely cured and back to full strength.

Because of today's obvious ethical constraints, it is difficult to study the nocebo response scientifically, but a few randomized, nocebo-controlled studies have been performed. Both British and Norwegian researchers have conducted experiments to determine whether there is a link between cell phone usage and complaints of headaches and migraines in cell phone users. In both studies, there was no difference in the severity of symptoms reported by the group exposed to real cell phone signals and by the group exposed to the sham signals.

Whether positive or negative, it is the expectation of a result that appears capable of triggering physiological reactions. Some researchers, however, have taken their studies even further, revealing that expectations are not the only cause of placebo and nocebo responses, but certain social cues can cause them as well. In 2008, Harvard Medical School scientist Ted Kaptchuk studied variations of the placebo response in volunteer subjects who suffered irritable bowel syndrome. One group was told they had been placed on a waiting list for a treatment trial; a second group received a placebo acupuncture treatment from someone who did not interact with the subjects; and a third received the same placebo treatment, but this time from a clinician who warmly engaged the subjects, asking them about their symptoms and talking about their conditions optimistically. The greatest improvement in health among all placebo groups? The subjects who received the placebo acupuncture treatment from the talkative and compassionate clinician.

The power of suggestion, expectation, or interaction with others to affect a person's sense of his or her own being is profound and suggests the self is not entirely circumscribed, but instead reaches beyond the fringes of the physical, beyond the confines of a single individual. In those cases where it's most difficult to know whether a person is truly a person anymore, a soul with a mind as well as a body, the presence of another person can be critical to making the determination. Just how much so was discovered only a few years ago. A traffic accident in July 2005 left a twenty-three-year-old British woman in a coma for several weeks. When she finally opened her eyes and began to cycle between sleep and wakefulness, she still showed no sign of awareness or any ability to respond to others. Failure to meet these two criteria is typical of patients in a persistent vegetative state, a peculiar twilight of the mind somewhere between coma and brain death, and usually with little chance of recovery. Traumatic injuries, such as motor vehicle accidents, strokes, and heart attacks, are the most frequent causes of a vegetative coma.

Brain trauma causes the severing of connections between neurons, but leaves the cells themselves intact, whereas a heart attack or stroke starves the brain of oxygen, causing cell death. Rarely do these latter patients recover enough brain function to truly wake up, since cell death is irreversible. Terri Schiavo's persistent vegetative state was due to massive oxygen deprivation following a heart attack. She became a cause célèbre for both conservatives and liberals in 2005 when her husband was given legal permission to let her die fifteen years after she lapsed into a coma. When her brain was autopsied, it revealed that she had suffered substantial and devastating decay.

For five months, scientists and doctors in Cambridge, England, studied the woman injured in the car accident using functional magnetic resonance imaging (fMRI). Random noises generated no change in her blood flow, but spoken sentences showed increased activity in the speech-recognition areas of her brain that was identical to the activity observed in the brains of healthy people. When the scientists asked the vegetative patient to imagine moving from room to room in her house, the neural pathways associated with autobiographical memory and motion began firing in her brain. When they asked her to imagine playing tennis, supplementary neurons in her premotor cortex were activated. With each question, the brain of the vegetative woman appeared indistinguishable from those of healthy volunteers responding to the same questions. The patient had already met the criteria for someone who was essentially brain dead, and yet the map of blood flow in her brain during the experiment indicated not only that she could hear, but that she could understand. In other words, she had consciousness. She had personhood.

David Alliance, a bioethicist at the University of Manchester, England, believes someone has personhood if they are "a creature capable of valuing its own existence." Using that threshold, the twenty-three-year-old woman would have failed the personhood test before the experiment. Only after someone—in this case a cognitive scientist with an fMRI machine—interacted with her and clearly elicited a reflective, nonautomatic response did she "pass." Months later, the woman did in fact awake from her twilight state and eventually returned home.

If interaction is key to personhood, then what to make of the thirty-two-year-old woman with hydranencephaly, a rare condition in which the cerebral hemispheres are missing? Treated by neurologist S. Allen Counter at Massachusetts General Hospital in Boston, the woman was born without frontal lobes where thought, reasoning, and awareness reside, and yet miraculously she had survived into adulthood. A quadriplegic, blind, insensate to touch, and lacking normal reflexes, the woman was missing both cerebral hemispheres and had just a clump of brain tissue attached to a small brain stem. All she could do on her own was swallow and breathe. Before she became Counter's patient, her doctors believed she had been in a persistent vegetative state since birth, but they wanted Counter to give them an independent opinion. Using a variety of noninvasive tools, he confirmed that the woman had no response to visual stimuli and no reaction when given mild electrical stimulations to her fingers and toes. But when Counter opened a child's music box in her room, she turned toward the sound and smiled, making noises of her own that convinced Counter that she was enjoying the music. Further tests of the woman's brain stem revealed that the neurons involved with hearing were normal. Counter wrote in a 2005 article in the *Boston Globe* that he "was so emotionally moved by her struggle for human definition through the single modality of hearing," that he went to a local electronics store and bought a cassette player and tapes of modern and classical music, which the woman enjoyed listening to until she died several years later.

Somewhere in that grubby remnant of gray matter was a person who came alive only because the outside world reached inside to find her. "Where does the mind stop and the rest of the world begin?" ask philosophers Andy Clark and David Chalmers in the first sentence of their paper "The Extended Mind." Their seminal essay was one of the first to propose a theory of "active externalism," in which parts of the environment become so fundamental to the way a person experiences the world that they can be viewed as extensions of the person's mind. The authors illustrate the concept with a thought experiment. Imagine two fictional people, Otto and Anna, who are both traveling to a museum independent of each other. Anna has memorized the directions and consults her mental map as she negotiates her way to the museum. Otto, on the other hand, suffers from Alzheimer's, and so he has written down the directions to the museum in a notebook. Instead of consulting his own memory, Otto consults his notebook as he travels to the museum. To Clark and Chalmers, Otto's notebook is no different from Anna's memory, and if memory is a crucial part of mind, then so is Otto's notebook. In his case, the notebook is part of his "extended" mind.

The idea is at least as old as Plato, who wrote a dialogue, *Phaedrus,* around 370 B.C. in which Socrates tells the tale of the Egyptian god Theuth who tried to get King Thamos to introduce to mankind Theuth's new invention, writing:

> "[I]t is an elixir of memory and wisdom that I have discovered," said the god.
> "You have invented an elixir not of memory, but of reminding," replied Thamus.

Today it is not just the tools of writing, but iPhones, iPads, search engines, and cell phones that constitute the mind's extended tools. Our cognition is unbounded; it reaches beyond the confines of the skull and into the world and makes the world, including its machines, an extension of ourselves. The mind, writes Clark, "is a leaky organ, forever escaping its 'natural' confines . . ." The mind is not, as Gilbert Ryle once said when he criticized mind/body dualists, a "ghost in the machine," but rather, a ghost outside the machine, something more than, or at least different from, the substance of the brain, something evanescent. Mind is not equivalent to brain. Rather, the difference between the inner and the outer, the mind and the environment, is relative. As individuals, we are the sum of all our relationships—with people and things, with space and time. We are hints and traces. We are unbounded. No wonder, then, that Sarkin felt so acutely alone. He was obsessively drawing and painting in part because he was obsessively reflecting on himself. In his overactive right hemisphere, the self-focus gave rise to a kind of constant questioning of himself, and a yearning to do something, say something—paint or draw something—in order to find relief from the relentless introspection. What he did not know, however, what he did not have yet, was a sense of connection to the world outside his head.

Chapter 26

BOLTFLASH

Art wasn't Sarkin's only means of expression. His boltflashes, the pages of rambling writing that often accompanied his drawings, were interior monologues, but they were also a means of reaching out. Interacting with people, making conversation and socializing, still felt awkward, and the boltflashes, even though they were one-sided, were Sarkin's way of connecting. He sent them to his friends, some of whom he hadn't seen since college or high school. He sent them to Tom Cruise, who had not yet made the movie of Sarkin's life. He sent them to perfect strangers with whom he had no connection whatsoever. If he liked something he read in the *New Yorker* magazine, say a review by Hilton Als, then Hilton Als was "boltflashed." Usually he did not hear back from these strangers, but occasionally, surprisingly, he did.

In 2000, Ira Glass, the host of *This American Life* on National Public Radio, called Sarkin. For months he'd been receiving boltflashes, mainly because Sarkin simply liked listening to the program while he painted and drew. Over the phone, Glass told Sarkin that he was planning to do a show called "Nobody's Family Is Going to Change."

The theme was based on a children's book of the same name, which was all about kids finally finding peace when they stop longing for their parents to be different. For the program, Glass wanted to ask if it was true: Does anyone's family ever change? The show was essentially three acts, the first was about a Jewish woman trying to find her way back to her estranged, born-again brother. The second was about two elderly sisters who never married and lived their whole lives together. The third act was "The Artist Formerly Known as Dr. Sarkin."

Glass interviewed the whole family, Kim, Jon, and the three kids. Curtis was twelve years old at the time, Robin almost nine, and Caroline six. Glass asked them about the strange things their father did, and they told him how he used to laugh maniacally at Mickey Mouse cartoons, the time he saved his leftover French fries in Florida intending to take them back to Massachusetts, and how he once wanted to skateboard with Robin and her friends. Another story was about the time when they were all getting ready to go to the beach and Sarkin picked up a basket of toys to put in the car.

"Mom says put the basket back in the garage," Curtis shouted to his father.

Jon promptly dumped all the toys onto the driveway and began to walk back into the garage with the basket.

"What are you doing, Dad?!"

"You said to put the basket back in the garage."

"*With* the toys!"

Sarkin had heard the words, of course. He'd strung them together one at a time, but when he did, their meaning dropped out.

"I'm so literal. It gets you into trouble," Sarkin told Glass.

Then there was the time he accidentally locked his youngest daughter, Caroline, out of the house in the middle of winter for a few, very cold, minutes. And the time he scribbled a phone number on one of Robin's drawings.

In the introduction to the story, Glass had asked, rhetorically, "What happens when you want your dad to change, and he wants to change too, but there's literally nothing that can be done to change him?"

In the middle of the NPR segment, Kim asked Curtis, gently, "Does he embarrass you?"

"Yes, in lots of different ways," he said.

When they all listened to the show on a Friday in August of 2000, Kim was devastated. Curtis had gone on to talk about how Jon was really a good dad, and that sometimes he couldn't help the things he said or did. That's why he didn't like to get mad at his father, he explained. But all of that was cut from the show when it aired. Kim knew, of course, how much the changes in Jon's behavior had affected the family's dynamics, and that when any of the kids got angry with him, he would become upset and despondent and feel like he was an incompetent father. She'd tried to help the children understand all of this by always talking to them openly and honestly. What she hadn't realized was how much her children still yearned for things to be different.

Around 2000, a friend of Sarkin's, also a painter, told him about a portrait class in town and suggested he try it. She told him that a live model's face or body was the only subject an artist could create and at the same time know exactly what it felt like to be the thing being drawn. A still life, a landscape, a ship at sea—these were objects forever outside the artist's being. Humanness was not. Sarkin was intrigued and joined a local life-drawing group, but every week he felt torn. The lessons created good structure, and he liked that, but he also dreaded going to the classes. Art was a solitary experience for him, always had been. Now, suddenly, he had to create in a crowd, in front of people who probably regarded him as downright strange. In his studio, he was safe and comfortable amid the chaos. No judgments. No expectations. And no possibility of getting into trouble, as he sometimes did at home when he bothered Kim or the kids by getting up in the middle of a conversation to draw or isolating himself in the bedroom with a sketch pad and a box of colored pencils. When he became upset, the studio was his safe haven and he would draw and draw, sometimes soothing himself by writing the word *shark* over and over and over, until it lost all meaning, all associations with fear, and was nothing more than a word. By that time, he would forget whatever it was that had made him so upset in the first place, and he could relax and draw again.

The students in the class brought their own canvases every week, but right before one session Sarkin realized he didn't have one handy, and he didn't have enough time to buy a new one. He looked around his studio for something, anything, to paint on, and uncovered an old gray metal shelf, about 3½ feet by 2 feet. He liked the coolness of it in his hand, the way parallel lines of small holes down each side reminded him of a giant strip of film. He primed the piece by covering it in acrylic and then, at the last minute, added the outlines of three boxes on it.

The class was from 7:00 p.m. until 9:30 p.m., and $8 covered the cost of the model. Sarkin carried the scrunched-up dollar bills in his pocket. He liked to arrive at the class a half hour late, after it had started, so that he wouldn't have to make small talk. The class was held in a small converted house and it was a cool, early autumn evening. With his cane in one hand and the heavy metal shelf in the other, he made an awkward entrance into the first-floor studio of the Rockport Art Association. Soft classical music played in the background as he propped the shelf on an easel and then dragged a folding wooden chair across the room to his station. Of the half dozen or so students in the class, only one other person had a stool. Everyone else stood, but Sarkin needed to sit. He had trouble balancing on his feet when forced to stand for long stretches. An egg timer ticked near the model, who held a position for half-hour increments before changing to a different stance. She was young and pretty, with long dark hair, and wore a colorful sweater vest over a long-sleeved white blouse.

Sarkin was dressed, as usual, in his artist's uniform: torn, paint-stained jeans and an old sweatshirt. He didn't talk to anyone, didn't exchange greetings, and didn't watch while the middle-aged man next to him opened his shiny oak box full of neatly arranged paints and began to draw an outline of the model's face on his canvas. Sarkin simply turned his backpack upside down and let dozens of pens and pencils clatter onto a nearby TV tray. Then he began to cover his metal canvas and the three boxes he'd drawn on it with white acrylic. When he was finished, he picked up a colored pencil and began to scratch, rapidly—and noisily—back and forth across the surface.

"Jon, what in the world are you doing now?" asked the middle-aged man, deep into his painting.

"Oh, I'm sorry, is this bothering you?" Sarkin responded, genuinely surprised.

"No, not me," the man answered. "I just thought it might be bothering someone else in the class."

The exchange unsettled Sarkin, made him feel even more self-conscious than he already was. He stood up and looked around, crossed over to the small stainless-steel sink on the other side of the room, and picked up a plastic, serrated steak knife. After covering the metal sheet again with white acrylic, Sarkin used the knife to carve the word *shark* into it, four times, in a vertical column. During the break, when the other artists put down their brushes, Sarkin kept at it, filling the metal plank again with swirls of color and then lightly drawing the outline of a face—a ghostly hint of the model, nothing more. Finally, when everyone else in the class was back at work, Sarkin stopped and stood up.

"Okay, I'm done," he announced to no one in particular.

It was not yet nine, but he stuffed his pencils and pens back into his knapsack anyway. A friend waited in the back of the room to give him a ride home.

"I'm really happy with this," he told her when they got outside. "It looks just like her."

The drawing didn't, not really, except in a very abstract way, but to Sarkin it did. The colors, swirls, the vague shape of a face—that was what he saw, and he felt comfortable acknowledging it. He knew he would never draw or paint like the others in his class, because he knew he would never see or feel the world in the same way they did. For the first time, that realization didn't bother him a bit. He had started the class that night nervous about being an outlier; he had left knowing he no longer cared, comfortable with his own vision.

"If you stare at a person long enough," Sarkin said, "eventually whatever you create will look like that person."

He still did things that had nothing to do with his art. He ate, he slept, he talked on the phone, watched TV, listened to music, went to the movies, and took out the garbage. But even then, he was always thinking about one thing, about art.

Chapter 27

WORKING IN THE DARK

O n January 30, 2001, Sarkin was in his studio, drawing, and had the radio tuned to National Public Radio and *Fresh Air*, hosted by Terry Gross. Her guest was Todd Feinberg, a professor of clinical psychiatry and neurology at Albert Einstein College of Medicine and the chief of the Yarmon Neurobehavior and Alzheimer's Disease Center at Beth Israel Medical Center in Manhattan. He had just published his first book, *Altered Egos: How the Brain Creates the Self*. The first thing Gross asked Feinberg was what he meant when he said he was looking for where the sense of self is located in the brain:

> "Where is your subjective sense of who you are, what you are and how you feel about yourself?"
>   Feinberg replied. "So I'm trying to find the neurological underpinnings of the inner 'I,' the self, the subjective self, the ego, the part of yourself that you feel is the most intimate aspect of your inner self."

Sarkin put down his pad and pens and listened. Gross asked Feinberg how studying brain injuries could help him figure this out.

"By studying patients who have brain damage, we find that many areas of the brain make a contribution to the self. So I set out—this is now fifteen years ago—through behavioral neurology and neuropsychiatry, to determine what areas of the brain contributed to the self. And we do that by doing an analysis of a particular area of the brain that is damaged and leads to a certain disorder of the self. Another area of the brain is damaged, this leads to a different disorder of the self. Trying to put all these areas together in the brain can help us understand how diverse areas contribute to the unified self that we experience as our inner 'I.'"

Sarkin hastily wrote Feinberg's name on a scrap of paper and looked him up on his computer. "You have to tell me what happened to me," Sarkin said to the neurologist by phone several days later.

After listening to this stranger tell him about his stroke and the changes he had undergone physically and mentally, and the compulsion to make art, Feinberg was hooked. He wanted to meet with Sarkin, maybe even scan his brain, but first he wanted to visit him and see what his life was like. Before the month was out, Feinberg drove up to Gloucester to spend the weekend observing and interviewing Sarkin and his family. Mostly, he watched while Sarkin painted and drew, meandered around Gloucester, ate dinner with his wife and three kids, and then went into his bedroom to draw again. At one point, Feinberg asked him about his art.

"It's like the Beatles song, from the *Sgt. Pepper's* album," Sarkin said. "You know, 'I'm fixing a hole where the rain gets in and stops my mind from wandering.'"

"Isn't it the other way around?" asked Feinberg.

Sarkin laughed. To a logical, linear mind, it's true, a hole in the head would seem to promote wandering, not prevent it. But for Sarkin, his art sealed him inside his own imagination, where he was happiest and most comfortable.

As impressed as he was with Sarkin's prolific output of drawings and paintings, Feinberg was particularly awed at his level of self-awareness. Brain-damaged patients suffering identity disorders are almost always unaware they have changed. They even sometimes resist the truth, like patients who are paralyzed on one side but instead of acknowledging they can't move their arm, they claim the arm isn't really theirs. And they will keep insisting this, despite being told over and over again, "Look, this is your left arm, it's connected to your left shoulder. It has the same sleeve, the same clothing, as your right arm."

The disconnect between brain and body has been studied in the laboratory. In 2008, Lorimer Moseley and his colleagues at the University of Oxford engaged volunteers in an experiment called the "rubber hand illusion." The illusion is induced by stroking a person's hand under the table, or someplace out of the person's sight, and at the same time stroking a rubber hand on top of the table and placed in a way that it looks to be part of the subject. The volunteer, of course, knows the rubber hand is a fake, and yet the trick is that the person perceives the "touch" in the rubber hand as if the rubber hand were his own. When both are stroked, the boundary between body and mind, what is real and what is not, becomes further blurred. The subject knows the rubber hand is not real, but suddenly it feels like it is. Moseley has even determined that the skin temperature of the real but hidden hand drops during the illusion, adding to the notion that the brain-body connection, at least in terms of our conception of ourselves, is tenuous at best.

In a very real way, patients with identity disorders are people with two minds. They suffer from a profound sense of self-alienation, while at the same time the truth of their own alienation escapes them. Their brains circumvent the strangeness of, say, a paralyzed arm, by creating a story, an excuse, if you will, for the recalcitrant limb. Feinberg's research into alien hand syndrome has even showed there is a peculiar gender-specific phenomenon associated with the disorder. Female patients with this particular identity disorder frequently mistake their left arm for their husband's, while male patients frequently mistake their left arm for their mother-in-law's.

Sarkin, Feinberg realized, was unlike these patients because he knew what had happened to his brain. He remembered the person he once was, and he understood the differences between his earlier life and the life he now was living. Sarkin was trying to find his way back to himself, too, but not through confabulation, through art.

Feinberg wanted to get a peek at Sarkin's brain and wrote him a prescription for a SPECT scan, which measures blood flow throughout the cranium. When Feinberg looked at the scans, he was taken aback. The stroke had damaged Sarkin's left hemisphere more than the right, but not only that, his entire neocortex was pockmarked with areas of damage.

When he called Sarkin, the neurologist told him the damage was difficult to assess because it was in so many different areas of his brain. Feinberg shared Sarkin's story with a colleague and friend, Martha Farah, the director of the Center for Cognitive Neuroscience at the University of Pennsylvania. She was particularly struck by the repetition in Sarkin's art. The repetitive behavior called "perseveration" is well known in psychiatric patients: someone who rings a doorbell and then keeps pushing the button; a letter that is written and then rewritten over and over, but never sent; a person who cannot change the subject of a conversation even when others try to. But Farah had never heard of perseveration in art. Most people who perseverate find the condition disabling, either at work or at home and often in social relationships. Perseveration as an artistic theme or style? No, she'd never seen or heard that before.

Feinberg told Sarkin that some of the more prominent areas of damage in his frontal lobes were most likely responsible for his general impulsivity and unpredictability in social situations, and perhaps even his obsessiveness. But why Sarkin felt a compulsion to make art, Feinberg said, he didn't know. When Sarkin left, he remembered something the neurologist had asked him. He wanted to know how his drawing style changed after the stroke. In the days that followed his visit with Feinberg, Sarkin sent him more than three dozen e-mails trying to answer that question, and more.

My art is largely "urborotic," an adjective I invented. Urboros (the symbol of a snake eating its own tail) is a visual metaphor for the stroke, for death as birth/rebirth. The urboros seems an apt description of my reality these days, and I try to incorporate this gestalt into my art. My reality (and hence my art) is not the left hemisphere dominant, logical, material, Western, Descartean, Newtonian, rational one of my pre-morbid self. It has been supplanted by a reality (surreality?) that is urborotic, right-brained, non-Euclidean, mystic/mysterian, Eastern, metaphysical, the antithesis of physical-based reality, illogical, irrational, intuitive, faith-based, transcendent and spiritual. My art is a metaphor of my emotional experience. An effect of my stroke is my inability to restrain my self-involvement. . . . Why do I write this crap? I think it is because my search for self (which, to me, is basically the same as a search for meaning) is my life now. And since my life is my art, my quest for self IS my art . . .

Chapter 28

THINGS SPEAK FOR THEMSELVES

I n April of 2003, right before his fiftieth birthday, Sarkin was re-
warded with his first New York art show. Again, his sister Jane
had a hand in setting it up. A friend of hers, Hamilton South, was
about to have a show at the Diane von Furstenberg Studio, a spacious
gallery and creative design lab in the Meatpacking district of Manhattan. Jon had recently acquired an art agent and Jane suggested that
she contact von Furstenberg's studio. The third week in April, Sarkin
and South opened a joint show at the downtown gallery. The evening
was mild, and a steady flow of people came and went all night long,
many of them purchasing pieces of Sarkin's art, which were priced
at a range from hundreds up to several thousand dollars each. Von
Furstenberg was present for the opening. So was the television executive Barry Diller, ABC broadcaster Diane Sawyer, and actress Meryl
Streep, whose husband, Don Gummer, is a sculptor. It was a heady
time. Kim and the kids attended the opening. So did Sarkin's mother,
stepfather, brother, and sister and many Gloucester and Rockport
friends.

Eight-year-old Caroline, in a long white dress, walked among the guests with a small notebook, shyly asking people their names and what they thought of the show. Jon held court, dressed in a sports jacket, his favorite bolo tie, and cowboy boots. At one point, an older woman approached him.

"Can I borrow you for a second?" she asked. "I just want to tell you what I think your art means."

Sarkin politely obliged and the woman shepherded him over to one of his larger pieces. He didn't pay much attention to her analysis and when the night was over couldn't remember what she'd said, except that she'd finally asked him what *he* thought his art meant.

"It doesn't mean anything," Sarkin said. "You want meaning? Go get the *Wall Street Journal.*"

Then he walked away. There was no intention to be rude or brusque. This was just how he was. He truly didn't think his art had any meaning—meaning was nothing more than the unconscious, almost automatic process of making the art, the immediacy of his sensations. Whatever was in his mind, no matter how fragmentary or random, came out in his art. Reason was not a part of the process. The stream-of-consciousness flow of words and images inextricably intertwined on paper or canvas was. Densely textured and detailed, his art drew observers in and encouraged them to follow a strand of thought or a winding line of color. Throughout the evening little red dots, indicating a drawing or painting had been sold, began to sprout beside the two dozen pieces of Sarkin's art hanging on the walls. When it was all over, Sarkin suddenly felt like his legs were going to give out. He had been standing, leaning on his cane really, the whole night and hadn't felt the least bit tired until everyone was gone. Now the gallery was empty, except for his family and a few friends. When his agent counted up the sales, Sarkin had earned about $20,000 in a little more than four hours.

By any measure, the night was a huge success. In the coming weeks jeweler David Yurman placed one of Sarkin's works in his Madison Avenue store window, and three other invitations for gallery shows were extended, two of them solo, one in Gloucester and the second at the SoHo studio of artist and photographer Jim Budman. A Manhattan money manager requested a painting for his office; a wealthy family wanted a portrait of their daughter; and Pingry, where Sarkin went to high school, asked if it could hold an art show and invited him to speak to its students. As his work became more widely known, so too did the story of his life, and articles appeared in the local press, in the *Boston Globe*, the *Boston Herald*, the *Ottawa Citizen*, and the *London Telegraph*. Bob Brown from ABC's *Medical Mysteries* wanted to include Sarkin in a show on brain injuries, and a British documentary team flew to Gloucester to film him.

Commissions were still rare and the money he earned from his art was highly unpredictable, but that didn't seem to matter much to Sarkin. In fact, he rarely set the prices himself, and more often than not let the buyer decide. One of his first big sales was to a man named Larry, a childhood friend of his brother, Richard. Larry had made a heap of money, and when he ran into Richard at a camp reunion, he asked how Jon was doing. Richard told him the saga—the surgery, the stroke, the long recovery, and now the art.

"Wow. That's amazing. I'd like to get a piece," Larry said.

When he called Sarkin, he asked him how much a work of his art would cost.

"Larry, here's what happens. I name a number and the client always says 'It's too high, how about this.' So it doesn't make sense to me to give you a number."

"Well, I didn't want to insult you . . . Okay, how about twenty thousand dollars?"

"I'm not insulted."

When he went home that night, he told Kim he felt like Larry was a Medici, a grand patron of the arts, and he was some struggling Renaissance painter. Sometimes, though, he said he felt like it was one giant scam, that he was taking money merely for doing something he couldn't help but do. Art was not lucrative enough for the family to live on, but Sarkin took it as seriously as any job he ever had in his life.

He drew a giant fish for Larry on a canvas measuring three feet by two feet and used up an entire box of 120 colored pencils just to fill in the fish's 20,000 individual scales. Larry loved it.

At a local show, a group of kids stood entranced in front of a drawing. When one of them asked how much it cost, he said, "What do you think?"

"Five dollars?"

"You're right!" Sarkin answered, and promptly sold the kids four drawings for twenty bucks, total.

He also bartered his art, once trading a painting for a handmade dining room table, and when the landlord of his studio admired a drawing, Sarkin exchanged it for two months' rent. The money was never steady, but it didn't bother Sarkin. "The income from my art is icing on the cake," he liked to say. But the work was hardly incidental. Sarkin knew this was his life now, that he couldn't do anything else—that he didn't *want* to do anything else—and so he worked hard at it. When he visited galleries, from the tiny Cape Ann museum to the Museum of Modern Art in New York City, he studied the masters, especially those who seemed possessed of a single theme or idea. It was all about practice and repetition.

"That's what you do. It's like the stories of those old blues guys who go to the crossroads and meet the devil and then when they come back suddenly they can play anything. Well, I met the devil, too. The real thing is, though, art is hard work. You don't go to the crossroads and the devil says, 'Now you can go and paint really well,' and suddenly you can. Drawing is hard. If people could get to a certain level of drawing proficiency like reading, or driving a car, everyone would do it: 'Hey, I see you passed your life-drawing test.' 'Yeah, it took me three tries.'"

In April of 2004, Sarkin took a train from Boston down to New Jersey as an invited lecturer at his old prep school, Pingry. He arrived at the school at 8:30 a.m., a half hour early, dressed in his blue blazer and loafers and wearing a tie. He graduated Pingry in 1971 thinking he might become a doctor or an architect. Now he was back as the guest of Miles Boyd, the head of the school's art department. It was all so crazy, Sarkin thought.

Standing in front of a dry-erase board and facing a dozen teenagers, he launched into a monologue two seconds after Boyd introduced him to the first class.

"What I do in my studio is I draw. I'm going to make believe there is a spy who comes into my studio. His name is Half Net man."

Sarkin drew a circle with a green marker and wrote, "The Adventures of Half Net," then "Voodoo King" with a blue marker, then the phrases "Detroit moon" and "Leonardo DiCaprio Da Vinci" followed by a series of spirals, numbers, and squiggly lines, and finally the words, "Could this be the end for Half Net?"

"I could stop now or go on for twenty-four hours," he told the class. "I have no preconceived idea of what I'm going to do. It's not like I'm setting out to draw a lighthouse. Doing stuff like this is different. So you have to turn off your brain to stuff that doesn't make sense. But imagine if the rules were different. Red means go and green means stop. It's hard to change your expectations . . . Imagine you had a blackboard a mile long and a pen that never ran out of ink and a lot of time on your hands—which I do. I could do this for a long time."

Picking up another colored pen, Sarkin drew a cartoon face with eight eyeballs, then put the cap back on the marker and announced, "Okay, I'm done. Now it's time for you to ask me questions."

A boy in the front row raised his hand.

"What's the hardest part for you?"

Sarkin answered without hesitating. He didn't mention the stroke or actually making art, but talked about time.

"What's hardest is always what's hardest in the here and now," he said, and here and now he was in front of a bunch of restless teenagers. "Usually I'm doing this on my own. I can do whatever I want. No one is judging me, saying, this guy's whacked, why isn't he doing something useful? Why isn't he a stockbroker? So the hardest part for me is I'm not normal. That part was turned off. So it's weird. But so what. The more energy I put into being self-conscious, the less energy I have."

He wasn't really talking to the kids now, but to himself. Another student raised her hand.

"Does this just come to your mind?" she asked, referring to the drawing on the board.

"Yeah. I put a lot of effort into not coming to mind," Sarkin answered with a smile. "Being undisciplined is very hard. If I gave you a piece of paper and said draw anything you want, that's kind of hard. Just drawing stuff out of your head is hard."

He was right, of course. Being in a state of "not coming to mind" was difficult for most people, but not for him—it was simply who he was.

"So here's what I did," he began to explain. "Half Net has no meaning at all. But then I made it like a comic book—the adventures with Half Net and the Voodoo King. Then I wrote Leonardo and everyone was thinking Da Vinci, right? So I put in DiCaprio, then Da Vinci. It's both. Whatever. If I'd just drawn the circles, it would have been like, whatever."

The drawings on the dry-erase board weren't really art, of course, more like samples of Sarkin's creative process.

"What's interesting," he continued, "is when I first started, the blackboard looked just like this."

In a flash, Sarkin erased everything he'd drawn. A couple of the students gasped.

"You have to start somewhere, so I started with a circle and it led to something else and something else."

Later, Boyd explained the reaction of the students when Sarkin erased his drawing. "To do good stuff you need insecurity—the drive to re-create again. A kid will draw something and save it and frame it. Jon just goes up there and erases it and then starts over."

Just before the end of the class, Sarkin got to the heart of the matter, the essence of who he was now, what made him so different, so unrecognizable from the man he was before his stroke.

"I have to fill up the space," he said. "There's a big blank space and when I see it I just have to fill it in."

He turned his back to the students, faced the board, and began drawing again.

"If I draw a line here, there's a lot of space, so I go over here."

He made a circle on the left and wrote the letter *A* on the right.

"So now I use red . . . then I keep going and I fill in the spaces. I might draw, write words, make a comic book, a face."

While Sarkin sketched on the board again, a young boy in the back spoke up.

"What got you into this way of thinking?"

Sarkin turned to the class, walked up to a redheaded student in the front, and picked up a drawing from her desk, a self-portrait in pastel.

"This looks like the Joker. Who didn't see that movie? You've got to go."

He didn't realize he'd insulted the poor girl; he was simply saying what he thought.

"I used to think like everyone else, and then I had a stroke and it really, really screwed everything up."

In one of the thousands of musings on art he sent to various friends via e-mail, Sarkin once wrote, "The true artist must operate in surreal reality, or he's not really being true, wholly, to his art. Most artists aren't willing to make this commitment. I don't blame 'em. This is NOT a pact to be made by the meek or partially committed. I don't think I'd have made this choice had it not been foisted on me. V. S. Naipaul: 'When a writer is born into a family, that is the end of the family.' Jon Sarkin: 'When an artist is truly born, that is the end of the person that was a person before he was an artist.' The true artist may have difficulty living in the world that most inhabit. He may be misunderstood, mistaken for being rude and uncommunicative. His business acumen may be questionable. His family relationships may suffer. He may experience bipolarity. What does he GAIN in this 'meta-Faustian' bargain? Well . . ."

The class was nearly over. Boyd asked if there are any more questions.

"Why do you use so many words?" asked the redheaded girl.

"Because there are so many out there," he answered. "If someone had brought me one hundred markers instead of these five, I'd have used as many as I could. I would just keep going."

The students in the class weren't quite sure what to make of the man, or his drawings, and so they kept circling back to the same question.

"Where do you get your ideas from?"

"I get ideas wherever you get your dreams from," Sarkin said, drawing a vertical Moebius strip on the board. "I don't care anymore about being an idiot. You got it made when you don't care . . . As they say in Latin, 'Res ipsa loquitur.' Things speak for themselves."

Chapter 29

Richard's Gift

K im, Jon, and the three kids had settled into a routine. Jon went to work every day at his studio; he had lunch at the Savory Skillet, maybe visited the library or the used record shop, or brought coffee to Ken Rieff, Gloucester's resident lawyer/filmmaker, in the storefront office across the street. He either rode the bus home at five or occasionally sat in with a few of his musician friends—not to play music, but to record some of his poetry.

Kim homeschooled Curtis, Robin, and Caroline, then drove them to their various after-school activities—art classes, dance—or to visit with friends. A stasis seemed to settle over them all, a kind of truce. No more expectations of Jon. For his part, he tried to involve himself more in his kids' lives, going with Kim to a play Curtis was in, or helping one of the girls with her homework. There was little that disturbed the pattern, until the fall of 2004.

On Tuesday evening, October 19, Jon's brother, Richard, boarded a twin-engine Gulfstream turboprop, en route from St. Louis to Kirksville, Missouri. The fifty-four-year-old pediatrician and professor at the University of Buffalo School of Medicine was headed for a teaching conference on humanism in medicine. Richard Sarkin was a familiar name in the Buffalo community. He had been an elementary and middle school science teacher before graduating from New York Medical College in 1977, and in addition to his hospital work, he was well known for designing programs that enabled other doctors to teach medical students how best to communicate with their patients.

The nineteen leather-padded passenger seats of the Corporate Airlines commuter plane were nearly full when it took off just after sunset. Sarkin and a colleague, Steve Miller, were both scheduled to give keynote addresses at the conference, and they sat across from each other in the front row, separated from the cockpit by nothing more than a curtain. The 210-mile flight north to Kirksville was scheduled to take just forty-seven minutes.

Pilot Kim Sasse and copilot Jonathan Palmer had spent nearly fifteen hours flying that day, hopscotching across the Midwest, most of it through rain and fog. Flight 5966 was their sixth, and final, flight of the day. The visibility was only a little more than 300 feet. Nearing their destination, the pilot had second thoughts about the heavy weather.

> Sasse: We're not getting in.
> Palmer: Jesus Christ. Go all this f— way. Well, let's try it.
> Sasse: Yeah, we'll try it . . . I don't want to . . . go all the way out here for nothing tonight.

Nine minutes and 33 seconds later, flight 5966 began its descent into Kirksville, dipping down into the thick clouds and mist.

> Sasse: . . . we're going into the crap. Look, ooh, it's so eerie and creepy . . . [I] get a suffocating feeling when I see that . . .

As the turboprop churned through the darkness, Palmer looked for the runway lights. So did Sasse, although it was contrary to the rules of the Federal Aviation Administration. The pilot's job was to watch his instruments and keep the plane on the right approach. At 500 feet, Sasse located the ground, but six seconds later, as the turboprop banked, there was a problem. The plane was two miles short of the runway.

Palmer: Trees.
Sasse:　No. Stop.

The plane's wings sliced through the forest canopy, beheading two 50-foot oak trees. The fuselage shredded.

Sasse:　Oh my God.
Palmer: Holy s——

The next morning, Jon Sarkin turned over in bed and grabbed the phone from his teenage son. Sarkin's brother-in-law, Martin O'Connor, was calling from New Jersey. It was sometime after 8:00 a.m., Wednesday, October 20. Kim was gearing up for the day's lessons; Curtis had already showered; and the girls were getting dressed.

"Hey, how ya doin'?" Sarkin said, settling his head back down on the pillow.

"Jane wants to talk to you," answered O'Connor.

A couple of seconds of silence. Then the voice of Sarkin's sister.

"Jon?"

"What's up?"

"Richard was in a plane crash. He's dead."

Jane had wanted to be the one to tell her brother. She told her husband, specifically, "I want to call Jon myself." She was only thirteen when their father died and as the youngest of three she had looked up to her two older brothers, idolized them actually. Even when Richard moved to Buffalo and Jon to Gloucester, the siblings remained close and often gathered at their mother's home, about twenty miles away from Jane and Martin and their two kids.

When Jane called her brother, she knew that thirteen of the fifteen people on board the small commuter plane that crashed in Missouri had been declared dead. Two unnamed passengers had miraculously survived. Somehow, she knew Richard wasn't one of them.

There was silence on the other end of the phone for what seemed forever.

"Oh my God, when did this happen?"

Kim was on the second floor of the house with the girls when she heard Curtis, who had been in the kitchen listening to his father, run upstairs.

"Mom, Dad's on the phone with Jane. It sounds really serious."

Kim raced downstairs. Maybe something was wrong with Jon's mother, Elaine, or perhaps his stepfather, Bill, who was in his late eighties.

"Okay," Sarkin finally said to his sister, still trying to get his bearings. "I'm going to go to Buffalo right away."

When he hung up the phone, he turned to Kim.

"My brother's dead. His plane crashed."

"No, no, no!" Kim screamed. She couldn't believe it. After all the family had been through, Elaine losing her first husband at an early age, Jon nearly losing his life. And now his brother was gone, killed in an airplane accident? It didn't seem possible.

Curtis shouted up to his sisters that Richard was dead, and the two girls flew down the stairs. Kim and the three kids collapsed on their knees, crying. Richard had been director of general pediatrics for newborn services at Women and Children's Hospital in Buffalo and also cared for children at a free clinic. When he wasn't practicing medicine, or teaching it, he coached soccer, and he was deeply involved with his kids. He was funny and wise, smart and athletic. When Kim and the children finally stood up, they wandered, separately, around the house, as if trying to find a way to shoulder their grief and disbelief. Kim turned on the TV and found CNN. There was an image of a smoldering plane on the screen. A reporter spoke:

"There's been a plane crash . . . a group of medical doctors . . ."

Sarkin heard the words as if from the opposite end of a long tunnel. He didn't want to look; he couldn't look. He remembered something a bus driver had once said to him. "When you're in hell, you don't stop to take snapshots." He walked by the TV on the way to the bathroom without moving his head.

Eventually the news reported what everyone in the family already knew—that it was all over in a matter of seconds. What remained of the shattered plane was caught in the broad branches of a thick oak, suspended 15 feet above the ground on the edge of a bean field. Many of the bodies were found still strapped in their seats, and the searing heat and fire made it difficult for the emergency personnel to recover the dead. It took a couple of days before what was left of Richard Sarkin was identified.

While the rest of the family walked around numb for the rest of the day, Sarkin made arrangements for them all to fly to Buffalo. This would have been ordinary before the stroke, but for the past fifteen years Kim was the one who planned the trips and made the plane reservations. Now, Sarkin didn't think twice about it. He just knew what it was he was supposed to do and he did it. For some reason the extreme stress of the situation focused him in a way he hadn't been able to in years, and a kind of calm overtook him that he found inexplicable but comforting.

Before they could fly to Buffalo, however, they needed to figure out what to do with their fifteen-year-old yellow Labrador retriever, Yolanda. She was going blind, had trouble walking, and was incontinent. Just that week they had all decided, as a family, that before their annual Thanksgiving vacation in the Caribbean they would probably have to put her down. She wouldn't be able to handle being boarded for a week, not in her condition. Now they had to leave town in less than twenty-four hours. Later that morning, all five of them, with Kim at the wheel and Yolanda in back, drove across town to the vet's office. They each hugged the dog good-bye, then the doctor gave her a shot and she closed her eyes. When she stopped breathing, a small puddle of urine formed around her belly. It was all too surreal, Sarkin thought. His brother was dead and his dog just peed all over the lino-leum floor as she was being put down. Later, they would mix Yolan-da's ashes with those of Ida, their black Lab who died in 1993, and bury the ashes at the base of a cherry tree in the backyard.

On the ride home from the vet, the Sarkins talked logistics. Who would take care of the cats, the need for Jon's suit to be pressed, what tie he would take with him. They were in survival mode, all of them, and for the first time in nearly two decades, Kim was able to lean on her husband for support. He was calm, coherent, and centered and knew exactly what to do. He packed his own suitcase: a somber tie, dress shoes, and the charcoal gray suit he always wore for weddings and funerals. He was numb, too, but he also felt good in a way. All the free-floating anxiety he felt every day, his disconnect with the world, had melted away. There was a focus in his life—a terrible focus, to be sure—but Richard's death had given Sarkin a purpose, at least for the moment, and it had nothing to do with him, or his stroke, or his loss of physical health, or his fractured mind, or even his art. It was primal and it was real, and it was about taking care of his family. They arrived in Buffalo at six o'clock that night, picked up a rental car, and drove the twenty miles out to Amherst, where Richard and his wife, Marcia, lived with their two teenage kids.

The next day, sitting at his brother's computer, Sarkin wrote a eu-logy that he told the family he wanted to deliver himself. His fingers seemed to glide over the keys, finding just the right letters, the right words. There was a presence, he felt—Richard, God—he didn't know, but it was real and it kept him moving forward.

The morning of the memorial service, Sarkin walked by one of the bedrooms in his brother's home and saw, out of the corner of his eye, his sister Jane and mother Elaine sitting on the edge of the bed talking.

"What are we going to do now?" Elaine said quietly, shaking her head.

Neither woman saw him, and he passed by without stopping.

There were limits to his ability to comfort, to offer solace beyond what he could express in words, or art, or by taking charge.

An overflow crowd of hundreds gathered in the auditorium of Albright-Knox Art Gallery in Buffalo at 11:00 a.m., Saturday, October 23, 2004. Richard's children, Alex and Jessica, read John Masefield's poem "Sea-Fever," and then Sarkin spoke.

"My brother was the best husband, father, brother, son, uncle, brother-in-law, son-in-law, nephew, cousin, friend, and neighbor. Ever. Most spouses don't express the way they feel about their wife or husband the way Rich did. I've never known someone to be that verbal about their love. Marcia was the same. He exemplified parenthood. Jessica and Alex will agree. He never missed a game. He was always home early to spend time with the kids. I have never seen a parent with more love for their children than my brother."

He had never felt this kind of pain before. It was searing, like the hand of God was pulling him through a fire. Horror cleanses, he thought to himself. It washes away all the inessential crap. Sarkin looked out into the ocean of faces, at his mother, who had now lost a husband and a son, at his wife and three kids, who had endured a kind of death in the family when he returned home from his stroke a different man. He was reminded of how, in the space of just a few seconds, the present can make ghosts of us all.

He wanted to end his eulogy the way Robert Kennedy had ended his impromptu speech in front of a crowd of mostly African Americans, just hours after the assassination of Martin Luther King Jr. in April 1968. Kennedy's death a few months later made a powerful impression on Sarkin. His father took him, along with his brother, to the train station in Elizabeth, New Jersey, not far from their home in Hillside, to watch the funeral train slowly pass by on its way down to Washington, D.C. For years afterward, Sarkin carried a small photo of Bobby Kennedy in his wallet. Without looking down at his notes, he ended his eulogy by quoting the words of Aeschylus that are etched on Robert Kennedy's tombstone at Arlington National Cemetery.

In our sleep, pain which cannot forget falls drop by drop upon the heart until, in our own despair, against our will, comes wisdom through the awful grace of God.

William James, the nineteenth-century philosopher and psychologist, once wrote that trying to understand consciousness was like trying to see the dark by suddenly turning on the light. The moment you try to capture it, you obliterate it. Somehow, though, Sarkin had caught a glimpse of his former self. In the five days following his brother's death, he saw, felt, breathed, and lived as he had before, in his former life. He had returned briefly to the Garden of Eden of self-awareness, and then just as abruptly, he was thrown back out again. Kim had seen it, too. For those few days after Richard's death, when the family gathered for the memorial service, her husband seemed almost normal. He wasn't needy; he wasn't eccentric. He was the man she had married two decades earlier.

Back home, however, Sarkin descended into a black, unforgiving place. *Now, I'm on my own again,* he thought to himself. He was depressed and irritable and looked for a way out through his art. Several weeks later, Kim, Jon, and the kids joined his mother and stepfather and sister and her family for their annual Thanksgiving vacation in Jamaica. For the first time in decades, Richard was not there and the reminders were painful. The entire family stayed at Half Moon Bay resort, and every day Sarkin walked out along the pier that extended from the beach a hundred yards or more, and sat and drew inside a gazebo. He had brought along his usual backpack full of art supplies—oil-based pastels, India ink, Sharpie pens, colored pencils and a piece of rolled-up canvas. On one of their last days in Jamaica, he unfurled a canvas on the floor of the gazebo and began to draw and write. He covered one corner with a fiery orange mist, another he drenched in black and blue. The wind picked up and began to howl, heaving the edges of the canvas into the air. He kept drawing, writing the familiar words and names: Lizard man, Rauschenberg, Warhol, Shakespeare, and stringing others together: shaman, mother,

Jane, dream. "It was the season of the fist" he wrote in the middle, just below "Bring it up, bring it on home." In the top left-hand corner, he wrote: "Phoebe Caulfield. This is for my sister. She's my favorite sister. She's my only sister." Sarkin had always liked J. D. Salinger's *The Catcher in the Rye,* and now the book resonated more than ever. Like Holden Caulfield, he had lost a brother and now he imagined his sister Jane as the catcher in the rye, saving him from plunging off the precipice of sanity.

Sarkin felt the usual drive, the hunger to put pen and paper together, but there was a different sort of power that he tapped, which made every stroke more meaningful than the one before it. His style was still random and spontaneous, but it was also now more purposeful. He knew at the end of the vacation that his piece was not yet finished, so he rolled it up and stuffed it into his backpack, and the whole time flying back he couldn't stop thinking about it. In Rockport that night, he scrounged around for old family photos and wedged them into his overflowing backpack and the next day, in his studio, unfurled the work-in-progress on the floor and picked up just where he'd left off. He glued pieces of photos to the canvas, pictures of Richard as a child, graduating from high school, and lying in a bathtub as an adult. Sarkin added a photo of himself when he was in his twenties, playing the guitar. Never before had he included family photos in his art. Never before had his art been so personal. The canvas seemed to become an extension of himself, a kind of reaching out to his brother, his sister, even himself.

Sarkin worked on the piece for nearly a month, which was unusual, and when he felt it was finally finished he began something new. After unrolling a blank, 4-by-6-foot piece of canvas, he thought of his brother again. The canvas was ripped, a corner of it torn off. First, Sarkin covered it with white house paint, then he took a bucket of green and threw it across the stiff fabric. For the next two weeks, he worked on the painting day and night, adding slashes of black and gray. He had always felt the urge to just get something out on paper was a kind of purging, but this drive was different. He wasn't simply ordering the world, structuring it; he was giving it new meaning. Instead of each stroke being an act of purgation, each was an affirmation. Making this piece wasn't enmeshing him further in his internal life; it was slowly freeing him from it.

When he felt satisfied with the final product, he had the canvas stretched and framed, even though one corner had been sliced away and was missing. He liked it that way. Damaged. It reminded Sarkin of those nineteenth-century marble grave markers—broken columns, books with blank pages. All the unfinished lives. That was Richard's life now—forever interrupted. It was his, too, of course, although he'd been able to pick the thread up again.

A local show was coming up, and he wanted to include this new abstract painting he'd just finished, but he couldn't think of a name.

"Into the Fog," suggested a friend.

"Perfect," said Sarkin.

He had been weighed down by the baggage of loss for so long. Now, in taking up his artist's life again, he realized that all the pieces were there even if they didn't always connect: his youth, his former life as a chiropractor, his stroke, his wife and children, Richard's death, even his growing success as an artist. Nothing would ever make sense, no matter how much he tried to assemble the pieces in his art. He would always be moving and changing, and he would always feel the motion more than most.

Chapter 30

INFINITY ON TRIAL

J on Sarkin is not a religious man. He didn't blame God for his stroke, nor did he credit him for his survival. But he believed everything happens for a reason—it's just that no one ever lets us in on what those reasons are. "Life sucks," he said. "That's a given. But art is a given too." So is the artist's life—always observing, always creating.

Back in the 1970s, after working his way through Carlos Castaneda's books about Mesoamerican shamanism, Sarkin had continued to read about mysticism, especially the Buddhist concept of samsara, or "wandering on." In the simplest of terms he understood it as the view that life was a journey, the continuous movement from birth to death to rebirth, with all the attendant confusion and angst in between. The Wheel of Suffering was not about place, but about process. We create our own worlds and we continually move in and out of them.

In the spring of 2006, Sarkin was fifty-three and had just been selected by the Boston art establishment as one of thirteen noted New England artists to be highlighted at the DeCordova Sculpture Park and Museum's prestigious annual show in Lincoln, Massachusetts. The curator of the DeCordova, Nick Capasso, didn't just want to bring

Sarkin's art to the museum. He wanted to bring Sarkin's studio to the museum. Capasso saw hints of Picasso in Sarkin's work, as well as the cartoons of R. Crumb, the gonzo art of Ralph Steadman, and the philosophical wordplay of nineteenth-century French Symbolists who combined text with images. Capasso was interested in something more than simply showing Sarkin's art. He wanted to take viewers into the process that led to the art—and what better way to represent that spontaneity and impulsivity than by presenting the art in the chaos of the artist's studio. So Capasso and an assistant drove up to Gloucester one day and, like careful sanitation workers, gently filled a dozen green plastic garbage bags with drawings, paints, and pencils and even some of the detritus strewn across the floor of Sarkin's studio. Back at the DeCordova, a team set up the exhibit, a 12-by-16-foot reproduction of Sarkin's workspace, smack in the middle of the gallery. They emptied the trash bags, tacked drawings to the walls, and scattered the rest along the floor.

On an unusually warm spring evening, Jon, Kim, and the three kids arrived at the DeCordova for the opening-night reception. In a few months, Curtis would be off to college, which thrilled Kim but also made her nervous. In the same way that art was her husband's life, the children were hers. She couldn't help but think about the future when all three would be gone. The only good thing about that, she realized, would be the grandchildren. Then she would have babies in her life again. Nothing stayed the same.

Kim picked up a drawing pad on a table in the middle of her husband's "studio" and smiled.

"I just got that book a week ago, Jon, and it's already full."

Sarkin wasn't paying attention. He was watching in amazement the crowd of people moving through this room-within-a-room.

"It's surreal. It's not my studio. But all these people are here walking through it, looking through my stuff. And I'm just watching them."

Sarkin knew he was a neurological freak, unable to see the world as a whole and unable to ignore it in all its infinite detail. He called the re-creation of his studio at the DeCordova *Infinity's Trial,* a play on the Bob Dylan lyric "infinity on trial." His explanation was simple: there's always more to come.

"I have to keep going. I can't stop. I want to have enough for infinity. For an infinity of shows, an infinity of art," he said.

Eight months after the DeCordova show, Sarkin was named a finalist for the $60,000 Wynn Newhouse award given annually to a professional artist with a disability. He didn't win, but it was yet another sign to Sarkin that he had arrived, had gotten a foothold in the art world. In the past, the judges for the award included Donna DeSalvo, curator at the Whitney Museum in Manhattan, Cheryl Brutvan, curator of the Museum of Fine Arts in Boston, and well-known New York City artists such as Chuck Close and Dorothea Rockburne. There was certainly nothing traditional about Sarkin's art. But it was expansive and it cut across all genres. Even as he became more established, more known, he pushed his own boundaries. He could work on a drawing with text and cartoon faces and cross-hatching one hour, and then the next paint a soft, lyrical abstract landscape. He began to work earnestly on a series of self-portraits, using large swaths of canvas that he hung on his basement walls.

Always it was about telling stories, about imagining other worlds using the bits and pieces of his tilted reality. So when he wasn't painting, he was writing. He found a cardboard Sylvania TV box someone in town had thrown out and he broke it down so that all four folds spread out like wings. Then he took a Sharpie pen and started a story, threaded with a variation on a Bob Dylan lyric:

Neon Jones woke up with that same sour taste. It reminded him of that time in Daphne's apartment on Kentucky Derby day. Then he revealed his dream He dreamt that evening's empire had returned into sand and vanished from his hand and left him blindly there to stand but still unsleeping. He dreamt that his weariness amazed him and that he was "branded" on his feet. He had no one to meet, and the ancient empty street he walked down was too much for him.

Before one of his local art openings, he tried to describe his artistic philosophy to a reporter from the *Gloucester Times*:

> The romantic ideal is this—we strive for something perfect in our lives, for the ideal of perfection, but whenever we think we get to it, we have to reach further to get to it. We feel if we work harder, try to be a better person, pray more, do better in life, that we'll get there. We don't. But we persevere. The common thread is the experience of the complete unreachability of perfection. This evening is about being in the moment.

And at night, when he couldn't sleep or he felt too tired to draw, he would let his mind go and write a poem:

The Best I Ever Had

fog has its advantages
they are few but advantages nonetheless
clinging to the rocks like barnacles unthinking and tenacious
money has its advantages too
betterment and never thinking about subway grates
the fog worsening
the fog clearing
tomorrows rain
today's bills
we revolve around the whispering fog lucidly
like some secret sequence of top-secrecy
counting out loud
reading aloud
foggy shoe leather
screaming like gulls
forgetting about our better nature
communicating disconnection
quietly joining the cold parade
consecrating influence

abdicating indifference
resolving completion
tinseling civilization
enduring confrontation
reading the paper
the slick coating of refusal
the cotton-ness of hardship
the hunger of mania
healthy choice
managerial insanity
a leader of men
a lead pipe of cornucopia
malfeasance
reasoning
the rock of gibraltar reveals nothing
i sacrifice everything
i call that a bargain

Chapter 31

THE LOQUACIOUS BRAIN

I n 2005, a year before the DeCordova exhibit, a friend of Sarkin's told him about a book she'd just read by the Harvard neurologist Alice Flaherty. In her memoir, *The Midnight Disease: The Drive to Write, Writer's Block and the Creative Brain*, Flaherty described a condition known as hypergraphia, or the compulsion to write, which she had suffered following the deaths of premature twin boys. Sarkin was fascinated. Here was someone who knew exactly what it was like to be overcome with creativity, with a torrent of ideas, and the obsessive need to put it all down on paper—or, in Flaherty's case, in notebooks, on Post-it notes, toilet paper, even her arms.

History, in fact, is replete with highly successful hypergraphics. Leonhard Euler, the eighteenth-century Swiss mathematician, produced eight hundred pages of writing every year of his adult life—and he lived to be seventy-six. Since 1911, seventy volumes of his *Opera Omnia* have been published, and there are still more to come. Psychologists have retrospectively identified hypergraphic writers, painters, and scientists, including Fyodor Dostoyevsky, Vincent van Gogh, Agatha Christie, and Stephen King.

The condition of hypergraphia is associated with the temporal lobes and often, as in the case of Dostoyevsky, and perhaps van Gogh, is related to epileptic seizures. Some schizophrenics, like the Unabomber, Ted Kacyznski, also exhibit hypergraphic tendencies, and drugs like cocaine can temporarily induce it. Robert Louis Stevenson, high on coke, wrote the 60,000-word *The Strange Case of Dr. Jekyll and Mr. Hyde* in just six days. Van Gogh created more than two thousand pieces of art in the last decade of his life and wrote detailed, six-page letters to his brother, Theo, at least three times a day.

The hypergraphic tendencies of some artists and writers, however, do not mean that all hypergraphic people produce works of art or literature, nor does it mean all prolific writers and artists are hypergraphic. The Reverend Robert W. Shields, a preacher and teacher from Dayton, Washington, who died in 2007 at the age of eighty-nine, wrote 37.5 million words during his lifetime. By comparison, Marcel Proust's seven-volume *Remembrance of Things Past*, one of the world's longest novels, is a shade under 1.5 million words, and the famous diaries of Samuel Pepys, just 1.25 million.

Shields, however, wasn't motivated by art and he didn't regard his recordkeeping as a way to reflect either on himself or his times. He simply needed to put everything down on paper. So for a quarter century he did, spending four hours every day on his porch, usually in his underwear and surrounded by six typewriters, recording the mostly insignificant events of his life in minute-by-minute detail—including the minutes he was recording his life.

April 19, 1994

12:55–1:05 I was at the keyboard of my IBM Wheelwriter making entries for my diary.

1:05–1:10 I lip-read Psalm 97

April 21, 1994

9:55–10:00 I fed the white cat, Ting, with canned chunk Tuna.

11:05–11:10 I served Grace ⅓ of the sausage and kept ⅔ for myself.

11:10–11:30 I ate the sausage while I went over the circulars that
   arrived in the mail.

April 29, 1994

1:49 We left the Safeway parking lot in the Plaza. It was sunny
   and brite.

1:52 We were at Ninth and Rose, Walla Walla, and it was 73
   degrees.

Shields recorded the prices of every item of food he and his wife
bought at the grocery store, described every one of his urinations
and bowel movements, and slept in two-hour increments so that he
could describe every dream. He even saved nasal hair and taped it
into his diary so that future scientists could someday study his DNA.
When Shields was interviewed by the *Seattle Times* in 1994, he offered
a reason why his ninety-four cartons of diaries, now located at Washington State University, might be of interest. "Maybe by looking into
someone's life at that depth, every minute of every day, they'll find out
something about all people. I don't know. No way to tell."

   After hearing Flaherty on public radio and then reading her book,
Sarkin knew he had to contact her. In the spring of 2005, he sat down
and sent her an e-mail.

   "I read your book. I think I suffer from that kind of thing too because I can't stop doing art."

   When Flaherty read the message she was both interested and
amused, not the least because this guy had just said he thought he was
hypergraphic, and he sent her, what, two sentences?! Surely, there was
more to this self-described hypergraphic than twenty words. They
exchanged several e-mails. Sarkin boltflashed her with his art and
poems. She invited him down to Harvard, where they talked for several hours and he attended one of her lectures at the medical school.
Finally, he brought her the MRI scans of his brain that Feinberg had
asked for a few years earlier. Like Feinberg, Flaherty was a neurologist, but she was also someone who knew intimately what it was like

to be Sarkin, at least in terms of a compulsion to write. Maybe she could make sense of his fractured soul.

"You have a beautiful brain," Flaherty said as she pondered the shadowy images of Sarkin's cerebrum on her computer screen.

Flaherty is wiry and fast-talking, but even when she's using arcane technical terms, she seems warm and approachable. As she scrolled through dozens of images, she kept up a running commentary, describing what she thought had probably happened to Sarkin.

"When you have a cerebellar bleed, you need decompression surgery to let the blood out, get the hematoma out," she said. "So you have to cut away tissue. It's remarkable how much of your brain is left. Most people just have a big old hole there."

Sitting in Flaherty's small, sun-splashed office, Sarkin looked at the neurologist expectantly.

"There may be a few dead cells in there," she continued, "but there is no swath of devastation."

There was damage, yes, and there was loss, but his brain also seemed to have been augmented. He was at once both lesser and greater than the sum of his parts, and Flaherty believed the changes were not so much anatomical as chemical.

"What I think is also very likely is that stress—from the stroke, the surgeries, the long hospitalization and then months of rehabilitation—has contributed to the changes in your brain," she said. "Heightened activity is an adaptive reflex in the face of stress. You're getting too much action in your brain trying to find a solution to your problems."

That activity, she explained, is especially elevated in the brain's right hemisphere when trying to solve a problem with creative insight. The left hemisphere and right hemisphere basically compete to figure things out, although the left usually dominates, relying on close associations that align with a general prototype. But if the analytic left hemisphere is all about linearity, searching for the textbook solution or dictionary definition, then the right hemisphere is like a thesaurus, gathering loose associations and creating patterns of meaning out of otherwise random information. The right hemisphere can do this be-cause many of its neurons have longer branches, so the cells can liter-ally reach farther to collect distant signals.

Most people who are looking for creative insight need to be in a state of relaxation where they can suppress the workmanlike, analytic activity of the left hemisphere—quiet the verbal chatter, if you will—in order to let their right hemispheres wander in search of a solution. Sarkin's stroke had unnaturally suppressed activity in his left hemi-sphere, so in a way his right hemisphere was picking up the slack, constantly in creative search mode.

Scientists have located the center of creative insight in a small fold of tissue on the surface of the right hemisphere called the anterior superior temporal gyrus, which has also been linked to the detection of literary themes and the interpretation of metaphors. All of this activity was being harnessed by Sarkin's brain to solve an essential problem: Who was he? If he did not know exactly how his brain had changed him, he nonetheless knew it had. Many hypergraphics, on the other hand, are unaware of the source of their compulsion. Their brains are working to solve a problem that remains hidden from them and silent.

This is what happens when people with primary progressive apha-sia (PPA) become hypergraphic. PPA is a neurodegenerative condition that slowly erodes speech and language while mimicking Alzheimer's. In rare cases, some PPA patients also develop extraordinary new ar-tistic or musical skills. Perhaps the most famous case of probable PPA, or a PPA-related illness, is the French composer Maurice Ravel, who died in 1937. He was known for building his compositions in small discrete, intensely detailed blocks. The great Russian composer Igor Stravinsky referred to Ravel as "the Swiss Watchmaker." In 1928, at the age of fifty-three, Ravel composed one of his most famous pieces, the hypnotic *Boléro,* which repeats its two alternating musical themes eight times over a span of 340 bars, each time increasing in volume and instrumentation. Although Ravel described *Boléro* as "an experiment in a very special and limited direction," neurologist Bruce Miller calls it a "classic exercise" in perseveration and compulsion.

Within two years of composing *Boléro*, the first signs of aphasia began to appear as spelling mistakes in Ravel's scores and letters. A taxi accident in 1932, in which he violently bumped his head, seemed to accelerate his symptoms. Both his speech and writing declined until he could no longer sign his name, and though he heard music in his head, he was unable to write it down or even read the music he had composed years earlier.

In 1937, Ravel underwent a series of X-rays after which doctors diagnosed him with "shrinking of the brain." Desperate, Ravel agreed to experimental surgery in December of that year. Doctors injected fluid into his brain in the hope of nourishing the receding gray matter, but Ravel never woke up. He died three days after Christmas at the age of sixty-two.

In a remarkable case of coincidence, sixty-five years after Ravel composed *Boléro*, Canadian Anne Adams painted *Unravelling Bolero*, unaware that she was suffering the same disease, primary progressive aphasia, which likely led to Ravel's death. Adams was a chemist and cell biologist and only dabbled in art until she was forty-six, when her son was critically injured in a car accident. Believing it would take months, if not years, for him to recover, Adams quit her research and teaching to stay at home and nurse her son. There, she took up her brushes and paints again and began to make art in earnest. After her son made a miraculously rapid recovery, she decided against returning to teaching and, instead, kept painting. Her early works were primarily realistic renderings of buildings and houses, but as she spent more and more time in her studio, her art became more abstract, more textured, and more colorful.

At the age of fifty-three—the same age Ravel composed *Boléro*—Adams painted her rendition of the composition, which she visualized in exceedingly minute detail: 340 individual rectangles, representing each bar of music, differing in size and color according to changes in volume and key. Seven years after creating *Unravelling Bolero*, at the age of sixty, Adams began to have trouble speaking and understanding others. By sixty-four, she was nearly mute. In 2007, at the age of sixty-seven, she was dead from progressive neurological decline.

Sarkin did not have to battle time or disease, and he had the luxury of extraordinary self-awareness. For that reason more than any other, he had long wanted an answer to the strange journey his soul had taken, to the mystery of who he was. He had sought out Feinberg and then Alice Flaherty, and as interested as he was now in Flaherty's perspective, he wasn't sure anymore how important it was to have answers.

"This is your cerebellum, with the unhealthy part on the left side," Flaherty told him, pointing to a shadowy mass in the lower portion of the picture of his skull. "And this is where the bleed was."

Sarkin wasn't really paying attention anymore. Instead, he gazed out the office window at the pattern of roof tiles on the building opposite the hospital. Flaherty was focused on the healthy-looking folds of gray matter in Sarkin's cerebral cortex. She didn't see much structural scarring. She could clearly see the cerebellar damage, but elsewhere, the white matter and the folds of gray tissue looked surprisingly healthy.

"Now does this mean your brain is the same as everybody else's? No," she said. "There are lots of changes in brain activity that aren't picked up in any of the scans I'm looking at. It's obviously doing a lot of interesting things that no one else's brain is doing."

In many ways, Flaherty said, Sarkin was a classic case of Waxman-Geschwind syndrome, a personality disorder characterized by excessive verbal output, an intensified mental life, and an obsessive preoccupation with detail.

"Everything is important," she explained. "There's a kind of 'stickiness' to things, to ideas, so that you can't break away from them."

Flaherty explained that multiple areas of the brain are involved in creativity: Before pen or brush is ever put to paper or canvas, the organizing and editing of ideas takes place in the frontal lobes. The ability to write is controlled by motor neurons in the cerebral cortex. The drive to write is controlled by the limbic system, which resides deep inside the middle of the brain, and the capacity for comprehension—that's the job of the temporal lobes, she said.

Flaherty believes it is the limbic system, the seat of our emotions and our most primitive drives, that kicks creativity into hyperdrive in those artists, writers, and thinkers who exhibit hypergraphia, and it is suffering and pain that pricks the limbic system to life. She knew how much Sarkin had been through, because he had shared so much of his life with her in their e-mail exchanges and conversations. And she knew the power of his almost manic interest in figuring out what all the suffering and pain meant. His art, his poems, his boltflashes were all reflections of his attempts to tease out some order from the chaos.

"Your temperament was always interested in ideas and your emotions were always up and down," Flaherty told Sarkin. "It just seems like they've all been intensified more than in the usual person."

"All the stroke did was amplify them, tweak them," said Sarkin in agreement.

He knew he was consumed with getting his thoughts and sensations down on paper, as if only then, looking at the colors and shapes and words, would it all come together into a pattern and make sense of his past and his present.

Some scientists, in fact, believe that the act of compulsive writing and art may be an attempt to understand or manage the deep complexity of existence. Change is ceaseless, but words on paper, images on a canvas are ways of halting the inevitable decay of present into past, of freezing the moment. The search, the attempt to order and stop time, is also an activity that can isolate a person. The more hypergraphics work to understand the meaning of life, the more their lives recede from the rest of humanity.

Sitting in Flaherty's cramped office at Mass General, Sarkin was getting anxious, eager to return to his studio. He was in the middle of a storm of self-portraits and the mammoth canvases, some 20 feet long, hung expectantly on the walls. An expansiveness had over-taken his art lately, and at the same time a concentration of energy. He was still doing his doodle art, his random drawings that were an amalgamation of words and images and color, but he was spending more time, too, on bigger pieces. The self-portraits were more realis-tic, less symbolic, as his focus sharpened. Maybe he was finding the edges of himself, filling in, at long last, the barren landscape of his bruised soul.

Chapter 32

FINDING TOMMY

My brain is split into a neuron war
Wondering and arguing what to do
Looking for a brain explorer, is it you?

The rhyme, from a man in Liverpool, England, appeared on Flaherty's desk not long after she met Jon Sarkin. When she opened the letter from Tommy McHugh, she thought, *Uh oh, here's a live one. Another case of hyperfluency.* He was a blue-collar construction worker and an ex-convict with a grammar school education, and now he had an unrelenting drive to draw, paint, talk, and write—and much of what he said and wrote was in rhyme.

Cup of tea, just for me, nice and sweet, just be neat.

Like Sarkin, McHugh's art had taken over his life—and his house. Born in 1940, he spent time in jail as a youth for assault and petty theft, and then, at the age of sixty-one, suffered a massive stroke. When he recovered, he often preferred speaking in rhymes, and he drew, painted, and sculpted for hours on end, often forgetting to eat or sleep. Whereas before the stroke he had been tough and uncompromising, after the stroke he was placid and mellow, even sentimental, and easily brought to tears. Although he had shown little interest in art for most of his life, it now consumed his every waking hour, and it scared him. Unable to cope with the dramatic changes in his personality, his wife, Jan, had left him. McHugh's small brick row home was now covered in art, from floor to ceiling—*including* the floor and ceiling.

He felt torn asunder, no longer the person he once was, and unsure of the person he had become. In so many ways he was just like Sarkin. He wrote to more than sixty doctors around the world, asking for their help, or at least an explanation. Flaherty was one of the only experts who wrote back. The two became pen pals.

McHugh sent Flaherty his medical records, and it didn't take long for the neurologist to realize that his brain had probably been injured in ways eerily similar to Sarkin's. McHugh, like Sarkin, was fixated on art and had trouble switching between activities, a sign of frontal lobe damage. But it was the changes in temporal lobe activity and the limbic system that impressed Flaherty again. The limbic system drives our emotions and links with the temporal lobes when we try to make sense of words. Suffering and stress, Flaherty believes, is what triggered the limbic systems of McHugh and Sarkin and sent streams of dopamine surging through their temporal lobes, driving them into furious artistic creativity. All because their brains were trying to make sense of their trauma, Sarkin and McHugh were in a kind of permanent and heightened state of "emotionality" due to their strokes.

When McHugh learned about Sarkin, the two began to exchange e-mails. McHugh was unable to travel to the United States because he was a convicted felon, and with changes in federal laws after 9/11, American immigration authorities would not allow him into the country. So Sarkin, along with Flaherty, flew to England in February 2007. McHugh met them at the airport. The short, stocky Liverpudlian with the coarse features and the paint-stained hands greeted Sarkin with a bear hug and a huge smile.

"Give me some love," Sarkin said.

For a week, McHugh barely stopped talking. As a friend drove the two men around Liverpool, McHugh read aloud the words on every street sign and often spoke in rhymes.

"I just want to say this to interest you. I want help to pull myself through."

"My brain is split in invisible wedges, leaving many Tommys on weak, crumbling edges."

"Tommy, Tommy, shut up," Sarkin would say, laughing. And McHugh would smile and laugh, too. He wasn't educated like Sarkin, had barely read a single book in his entire life, and had never really left Liverpool, except to do time in jail. At a café in the city, Sarkin ordered a double espresso.

"What's that?" McHugh asked.

When Sarkin explained, McHugh ordered one, too, and then was amazed at the tiny cup it came in. But there was also so much about this man that Sarkin related to. Talking in his apartment, McHugh often picked up a pipe and spun doughnut-shaped washers along its length. He would do it over and over, almost like an autistic person, absorbed in the motion of the washers as they spun around and down the piece of steel.

"I like to watch things spin," he told Sarkin. "It's calming."

"I'm the same way. I like to write the same word over and over. The same thing, over and over."

Sarkin said meeting McHugh was like being in a *Twilight Zone* episode where you travel to another planet and meet your twin. He told McHugh he was obnoxious, like a twelve-year-old, a bad boy and uncontrollable—just like him.

I can't believe how much this guy is like me, Sarkin thought. It was, he told McHugh, "very validating, very sobering."

"You're the real deal, Tommy. We're the same. We see things other people don't, and we're crazy, but we make sense. End of story."

For a week they drew and painted in tandem, whether sitting on the train, each with a pad, sharing pens and pencils, or side by side in McHugh's basement, or on the floor, at opposite sides of a six-foot canvas, painting toward each other.

"I'm whacked, yes?" said Sarkin, cackling.

"I'm happily whacked," said McHugh. "My mind is like a volcano, exploding with bubbles, and each bubble contains a million other bubbles, and each of those bubbles contains a million other bubbles," said McHugh. "I can't stop it."

"I can't *not* do it," said Sarkin. "I believe in God. And I hope to God I never stop painting. It's not quite as important as breathing, but it's up there."

How we feel about ourselves—physically, emotionally, spiritually—is not a solitary process of self-reflection, but has everything to do with our expectations and interactions with others. Sarkin and McHugh had come from very different places, with very different backgrounds, and yet they ended up in the same unusual place. If Sarkin had been the "test" case for this strange artistic life, then McHugh was confirmation of its reality, and that helped Sarkin feel more fully alive. After meeting Tommy McHugh, he felt comfortable for the first time with who he was now. He realized his identity was not a mirage, nor was it simply wrapped in the surreality of his interior life—it was out there in the world, in Tommy McHugh. His message in a bottle had been found.

Chapter 33

Best Friends

K im and Jon had held their marriage together through hope, faith, memory, and their commitment to their three kids. They had come farther together than either of them thought they would in the first few years after the stroke. But there was so much about Jon that was still hard for Kim to understand—until she found herself struggling, just like Jon had twenty years earlier.

On Tuesday afternoon, October 28, 2008, Kim and Caroline drove through a soft, chilly rain to the North Shore Mall in Peabody, about fifteen miles from home. Robin was taking an extra literature class nearby, so they were just killing time. Their destination was Williams-Sonoma, the home furnishings and gourmet cookware store. Caroline had finished her home lessons for the day, and Kim looked forward to getting out of the house for a couple of hours. She had been anxious, mulling over a decision about Caroline taking some outside classes. On top of that, she was worried about their four-year-old cat, Pepper, who was missing. She had disappeared once before, for ten days, and come back musty, dirty, and hoarse, as if she'd been trapped some-where, crying for hours to be freed. This time she'd already been gone for two weeks and all Kim could think about was Pepper starving to death in some shed.

Caroline helped to lift her mood. They had an errand to do to-gether and as they entered Williams-Sonoma she cracked a joke that sent her mother into a spasm of laughter. Just inside the door, Kim suddenly felt a kind of electrical charge surge through her head. She stopped, reaching out a hand to a nearby display table to steady her-self.

"Jesus Christ!" she yelled.

The room was spinning hideously. Her knees buckled, and with her other hand she grabbed the side of a bookshelf. The next ten min-utes were the longest in her life. As the room kept turning, Caroline held her arm and guided her mother out of the store and to a bench in the middle of the mall.

"I don't know what to do," she said, starting to cry. "This is the worst thing I've ever felt."

For ten years, Kim had experienced sporadic dizzy spells. They were disturbing, but they lasted just two or three seconds—no spin-ning rooms, not like this. Afterward she usually had a bad headache. This one, though, this one was a hundred times worse. Maybe she'd had a stroke or a seizure. She didn't want to alarm Caroline, who was already afraid her mother might collapse.

"Let's try walking, and see what happens," Kim said, tentatively. "But hold onto me."

Clutching her mother, Caroline guided Kim another ten or twenty yards, as far as the next bench, then gently sat her down. Kim called her doctor.

"I think I've had a seizure," she told the receptionist on the other end of the line.

"No you didn't."

That should have been reassuring, but it was slightly annoying. What did this woman know about what she was experiencing? The receptionist told her to come to the office immediately, so Kim called her half-brother Adam, who was home from college, to see if he could drive them. When Adam arrived, Kim was still shaking, afraid to get up because she thought she might fall down. He and Caroline braced Kim on either side and guided her slowly out of the mall. First, they picked up Robin after her class ended, then Adam drove them to the doctor's office. Kim's blood pressure was extremely high, but her EKG was normal. The physician gave her a prescription for Valium and suggested she see a neurologist. When she got back home, Kim called Jon at his studio.

"Why didn't you call me first?" he asked.

"I don't know. I guess I needed someone to pick me up and get me home right away."

Kim lay down on the bed and waited for Jon, who immediately took the next bus home. Nauseous, she couldn't close her eyes because that made the room spin faster. When Jon came in, he sat down next to Kim.

"What's happening to me?" she asked him.

Jon never doubted what Kim was going through. He had lived through all of it before—nearly twenty years to the day, when his head "exploded" out on the Cape Ann golf course. How bizarre, all this time later, that Kim should go through something so similar. That night he e-mailed Alice Flaherty. "I need a favor," he wrote. "Can you see Kim?"

The next few days Kim was afraid to go to bed at night, thinking she might die. Her heart pounded with anxiety and her neck ached. Jon talked her through, reassuring her that this wasn't something that was going to kill her.

"It's not a tumor, Kim," he told her. "It's probably something chronic and this is just a really, really bad episode. It will get better, I promise."

Within a few days, the ringing started. She couldn't believe it. Just like Jon. Was this some sort of twisted joke? The sound was like crickets; it was loud but luckily nothing like the sounds Jon heard years earlier.

Flaherty was able to see Kim a week later at Massachusetts General Hospital and gave her a thorough neurological exam, just like Jon's doctors had done two decades earlier. Everything was normal, just like it had been with Jon. Flaherty sent Kim to a vestibular specialist, but he couldn't find anything wrong, either. Kim was given a prescription for an antianxiety medication and Celexa, an antidepressant. There was probably no coincidence that the episodes of vertigo in the past had almost always occurred at times of great stress in her life.

For Kim, thankfully, the vertigo and the tinnitus eventually subsided, though it took many weeks. After a month, she was functional, but still extremely uncomfortable, unable to read or work on the computer. When she drove Caroline to dance class, instead of waiting inside with the other mothers, she sat in her car, pushed the seat back, curled up, and took a nap. At home, Robin helped cook, Caroline did the laundry, and Jon took the dogs out and kept the kitchen clean. By Christmas things were better, but the exhaustion and the fear of another horrible attack of vertigo lingered. Kim was eating a peanut butter cracker one day, moving her jaw around, and suddenly felt that ping again, a surge of electricity rushed through her head, and then a few seconds of dizziness.

Aha, she thought. That was it. Anytime she had the vertigo, she had neck or jaw tension. In her sleep, she clenched her teeth and slammed them together. In her mind, they were all related. Her dentist told her she had a malocclusion and realigned her teeth. The fact that she probably had a temporomandibular joint disorder—TMJ— was another bizarre coincidence. This was Jon's specialty when he was a chiropractor. He would never have wished this upon her, but in a strange way the fact that he had already been there, that he knew exactly what she was going through, made him feel useful to her in a way he hadn't for a very long time. Kim had never doubted Jon's agony all those years ago, but she knew exquisitely now what that agony was like. Thank God she had Jon, she thought.

Chapter 34

UNBOUND

J im Amadeo, the creative director of Mullen, an ad agency outside
Boston, was looking for inspiration one day in 2008. He turned
on the TV and began to watch an ABC *Medical Mysteries* special. The
final ten-minute segment was called "When Stroke Inspires Creativity," and it was all about Sarkin, Tommy McHugh, and Alice Flaherty.
Amadeo had never heard of the concept of spontaneous creative output, or of the three people profiled, but when he heard that Sarkin lived
in Massachusetts, he got up, went to his computer, and Googled the
artist's name. Then he bookmarked the page. Sarkin, he felt, embodied
just the kind of creative energy and free spirit the ad agency liked to
promote about itself. Maybe there was something the agency could
do with him down the line. In June of 2009, the opportunity arrived.
The agency was moving from the suburbs to the financial district in
downtown Boston. Walking around their new digs, Amadeo realized
that the long wall facing the coffee bar on the tenth floor would be a
perfect canvas for someone like Sarkin. He sent him an e-mail and
then the two spoke by phone. Amadeo told Sarkin about the space. It
was 26 feet long and 5 feet high, and it was all his. He could do with

it whatever he wanted. There was a one-word theme the agency often used to describe itself, and he offered it as inspiration:

"Unbound," Amadeo said.

"Do you know what you're getting into?" Sarkin replied.

This was the deal: Mullen would pay Sarkin $5,000 and would foot the bill for all his supplies, as well as meals and a room at the Hilton while he worked on the mural. If he wanted to travel to and from Gloucester every day, that was fine, too. The agency would pay for a driver. Sarkin felt like a king. Then he spent like one, filling a grocery cart full of $2,500 worth of art supplies: spray paint, oil- and water-based pastels, acrylic and oil paints, gauche, watercolor, Wite-Out, fixative, glue, and crayons. Oh yes, and a pair of taillights from an auto parts store.

He had a week, and he started on Monday, June 8, 2009. A sedan came for him early each morning, along with someone from the agency, and as the car sped down to Boston and crossed over the Tobin Bridge into the city, Sarkin stopped talking and got into the zone. When he arrived on the tenth floor, there were already a few people sitting at the tables in front of the coffee bar. He began by writing "The Adventures of Unbound Man" across the top of the long white wall, then added strings of words that streamed from one side all the way over to the other. Each hour, each day, he added and subtracted, layering colors over the words and words over the colors. People came and went, they looked and they talked, and Sarkin asked them questions:

"Who's your favorite musician?"

"Van Morrison," said one of the Mullen employees.

And by the end of the day, Van Morrison's portrait was part of the mural.

A secretary said she thought there should be women in the mural. A little while later, there was Aretha Franklin on one side, Amelia Earhart on the other.

Someone said they liked the Silver Surfer. The next day comic book cutouts of the Silver Surfer appeared on the wall. For Sarkin it was a whole new experience. He wasn't holed up in his basement studio; he was making art out in the public and the public was contributing. He fed off it, encouraged it, and the mural grew.

One morning, someone at the coffee bar handed him a photo of Abe Vigoda, from the old TV sitcom *Barney Miller*. A friend of Sarkin's, who was visiting, suggested a favorite quote from Jack Kerouac: "Fame is like old papers blowing down Main Street."

Other people brought in art prints, pictures of boats, and one of Jesus. An executive mentioned Willy Wonka. That night Jon told Kim he was doing "full-Ninja art." Total involvement. Images and words came and went. He reproduced the mouth of Marlon Brando screaming "Stella!" and juxtaposed it with the mouth of Aretha Franklin singing "Respect." He riffed on the last paragraph of F. Scott Fitzgerald's *The Great Gatsby*:

> He had a come a long way to here and his dream was so close that he could hardly fail to grasp it. He didn't realize that that dream had passed his ass good-bye, and had done reconsidered some place back in the yawning maw of ineffability, back home in Hillside, New Jersey, where the plainclothes inventioneers skateboarded under the boardwalk out of the sun. Most of the large haciendas had been abandoned to the dust bowl.

He finished the mural on the eighth day, and he rested by going across the street to a cigar bar and lighting up a stogie. He arrived home about one in the afternoon, took a nap, and didn't wake up until midnight. Mullen celebrated the completion of the mural with an art opening, and Kim and the kids came, awed at the scope and breadth of the work. The whole process had been more art performance, or event, than just art. Someone tweeted about it while Sarkin painted; another person filmed the mural's creation. For Sarkin, it was a kind of culmination. The massive work had bits and pieces of all the art he had experimented with over the years, from collage to portrait to cartoons to words and puns, photo scraps and stencils—and of course, taillights. It was also permanent. From the creative director on down, the people at the agency loved it and loved Sarkin, and they highlighted the mural and Jon on their website.

A year after the Mullen mural, Sarkin was handed another remarkable opportunity: an album cover for the alternative rock and pop band Guster. Guster's three original members met in 1991 as freshmen at Tufts University in Medford, Massachusetts, a small suburb just outside Boston. Adam Gardner, Ryan Miller, and Brian Rosenworcel stayed local after graduating, playing gigs up and down the East Coast and touring relentlessly, especially on the college circuit, gradually developing a small but fervent fan base.

Rosenworcel, the drummer, learned about Sarkin through his wife, Megan Poe, a psychiatrist who grew up near Gloucester and met Sarkin in 2008. She purchased several pieces of his art, and Rosenworcel fell in love with their whimsy and eccentricity.

In 2010, Guster was getting set to release its sixth album, *Easy Wonderful*, and the band was mulling over ideas for the album cover. The previous five albums had all been produced by the same man in Boston, but Guster had recently signed with a new label, Universal Republic Records, and was looking to go in a different direction. Most album covers are designed and generated via computer, but Rosenworcel had a better idea. "In my dreams, can we get Jon?" he asked the band's manager, Dalton Sim.

When he was contacted, Sarkin was thrilled and asked Rosenworcel to send him some of the band's music. Guster sent the songs from *Easy Wonderful*. "I'll see if it inspires me," said Sarkin.

Within a week, a New York contact for Guster's new label started receiving strange pieces of art in the mail—a storm of art, really—until an entire room at the New York offices of Universal Republic started to fill up with Sarkin's drawings, paintings, and collages. The band loved it all, and eventually settled on a cover that was essentially an intense color wheel with Guster's name spelled out in vibrant yellow against a lavender banner, with each letter covered in Sarkin's characteristic cactus spikes.

The band also needed a music video to accompany the October 2010 release of *Easy Wonderful*. A half dozen directors pitched story ideas. When Chad Carlberg, a Gloucester filmmaker, included Sarkin's art in his story line, the band and the music label jumped: this was a unique chance to tie the packaging of the album in with the music video.

For a single, intense week in August, Sarkin gathered with the band members and a film crew in a theater at Gordon College in Wenham, Massachusetts. The final video accompanying Guster's upbeat, melodic single "Do You Love Me" is as much about visual art—and Sarkin—as it is about Guster and music. It opens with flashes of Sarkin's artwork. Then the band appears dressed in solid-white long underwear—blank slates to the colorful cacophony of Sarkin's art that frames the stage. Nothing in the video is computer generated, although some of the art comes to life as animations. Stars and hearts, all drawn by Sarkin, are lowered onto the stage behind the musicians; Rosenworcel's bass drum becomes a giant, moving color wheel, the same one that appears on the cover of *Easy Wonderful*. Then a table-top-size color wheel is spun onto the stage by the band members. A guitar is replaced with a Sarkin rendition of a guitar; the words the band is singing appear overhead, in Sarkin's handwriting; a Sarkin-drawn heart pulses in the chest of one of the musicians, then finally, toward the end of the video, with art swirling around Guster, Sarkin himself makes an appearance. Dressed in a white bathrobe, he starts to paint the band members—splashing their long johns with yellow, orange, and blue—and when the curtain closes at the end, it is a curtain that has been drawn by Sarkin.

During the filming, Kim and the kids visited the theater and watched an ebullient Jon at work. This was something much more than the interactive Mullen mural. This was music and art being made together, Sarkin and the band engaged in an act of collaboration and mutual creation.

"It was my world brought to life," said Sarkin. "My art was animated. And I was just one of the characters."

The "Do You Love Me" video was chosen by iTunes to be one of its music videos of the week in September, and when Guster launched its fall tour in October, the *Easy Wonderful* T-shirts sold at the concerts were all covered with Sarkin colors, images, and writing.

For Sarkin, the collaboration was an affirmation of his identity as an artist. For Kim, it was another sign of Jon not just returning to the world, but embracing it. Sarkin had learned from Tommy McHugh that he was not alone, and from the Mullen mural and his work with Guster that interaction and collaboration could take his art to another level, that he needed others—their ideas, their caring, their energy— to help him make sense of himself.

Chapter 35

FLAWED WORDS,
STUBBORN SOUNDS

The imperfect is our paradise.
Note that, in this bitterness, delight . . .
Lies in flawed words and stubborn sounds.

—Wallace Stevens, from "The Poems of Our Climate"

W hen neurosurgeon Wilder Penfield was operating on epileptic patients in the 1930s, he occasionally took time to
perform an experiment. In one, he used his electrodes to temporarily prevent his subject from being able to speak or to understand
speech. A second before his assistant showed the patient a picture of
a butterfly, Penfield applied the electrode to the speech cortex of the
patient's brain and then asked him to say what he saw. Blocked from
speaking, the man remained silent for a few moments, then snapped
his fingers, trying desperately to circumvent his inability to communicate. When Penfield withdrew the electrode, the patient described
his experience.

"Now I can talk," he said. "Butterfly. I couldn't get that word *butterfly*, so I tried to get the word *moth*!"

Even though his speech was blocked, the man clearly understood the meaning of the picture and made a conscious effort to "get" the right word. Even when he couldn't, Penfield later wrote, "he turned back for a second time, not to the verbal center of his left hemisphere, but to the interpretive mechanism in his right, which was well away from the interfering effect of the electrical current." There, in the folds of tissue of the man's right hemisphere, he found the closest thing to a butterfly, a moth. "He must then have presented that to the speech mechanism, only to draw another blank," Penfield wrote.

The patient saw the picture and clearly understood what it was; he just couldn't say the word. For Penfield, the result of the experiment spurred him to conclude that something else—a soul, a mind— was directing brain function, trying to find a way around the brain's failures, in order to make sense of the experience. More than vision, more than language or even speech, making sense is what keeps us going, what defines us at our cores.

In 1981, anthropologist Arthur C. Custance summarized the history of "soul searching." He said with Descartes, "psychology lost its soul and found its mind." With the British empiricists, the "soul lost its mind and found its consciousness," and with the behaviorists, the "soul lost its consciousness and found its reflexes." If he hadn't died in 1985, Custance might have added a fourth, contemporary version, courtesy of the cognitive neuroscientists, for whom the soul lost its reflexes but found its neural correlates.

Nothing, however, remains fixed. Everything is local and always moving. The whereabouts of the self shift from lobe to lobe, hemisphere to hemisphere; it "wanders on," fragile and fitful, the sum of more possibilities than there are stars in the heavens. We are, all of us, an amalgam of associations—what we see and hear, smell, touch and taste—and the memories and emotions they conjure. Our ability to understand the world is limited by our very humanness. We are fragments, our sensations mere splinters of reality. Like standing outside one's home at night and peering through the window, all that we see of the world is what passes briefly into the light of that small frame. The rest remains in shadow.

From these meager ingredients our brains build our identity without our ever noticing. Sarkin was like that man trying to see the darkness by turning on the light. The truth of who he was kept escaping his grasp. His soul, he knew now, would always be in transit, because the world would always feel fragmentary. One moment would never lead to the next. Instead, each was an island in time, isolated from everything except his own imagination. Like a spectral Robinson Crusoe, he would forever be shipwrecked on the rock of his own consciousness.

On a bright, sun-splashed August day, the operatic cries of Gloucester's seagulls floated down the basement steps into Sarkin's studio. The walls were covered with large self-portraits. Random drawings and paintings lay in heaps on makeshift shelves, and shards of colored pens and pencils were flung across the floor. Sarkin searched for a tube of white acrylic among the plastic cups, scraps of paper, and pens and pencils. Finally, he located one, but it had already been squeezed dry. He laid a piece of wood across the bottom of the tube and pounded on it with his fist. A single short splurch of air and a few wet molecules of paint sputtered out, nothing more. Now on his hands and knees, Sarkin continued to scrounge until his fingers uncovered another slender tube. Not paint—sunscreen. It was white, he thought to himself. Why shouldn't it work? He even enjoyed a humorous thought: if it doesn't work, at least the colors won't fade because they'll be UV-protected. He tried to work the sunscreen into the portrait but finally gave up.

It's too nice to be indoors, he thought. So he switched gears, rolled up a 12-foot-long seascape he had been working on earlier, tucked it under his arm, and headed down to the beach. He wanted to see what would happen to the chalky colors if he bathed the canvas in seawater and rubbed sand into its seams. The five-minute walk twisted and turned through pieces of Sarkin's past. On Main Street was the Blackburn tavern, where he first met Kim. On the corner, the American Legion Hall where Kim celebrated her fortieth birthday and Jon sang with the band. Down to Rogers Street and the empty lot where his chiropractic office once stood. Past the Capt. Vince seafood company with the curious sign, "Live Lobster Entrance."

He slipped between the Chamber of Commerce and the abandoned Bird's Eye factory to the edge of the beach. Kicking off his old Docksiders, he leaned forward, letting his feet find purchase in the sand. At the edge of the water, he tossed away his cane and stumbled into the gently lapping waves. When he unfurled the painting, it billowed just for a moment, then settled on the surface. The seascape mingled with the sea. Art with life. Never one without the other. He dragged the soggy canvas from the water, across the sand to a rusted cyclone fence behind the boarded-up factory and hooked it to the top. Then he stood back to admire his creation. Just then, three young kids walked by.

"What is it?" one of them asked.

"It looks like the ocean," one of the others said.

"It looks like the sky," said the third.

They kept walking. Sarkin liked their answers. He considered leaving the large canvas just where it was, sagging on the ancient fence, facing out to sea.

"This is where it wants to be," he said. "It doesn't want to be in a gallery."

He was ready to leave it there, except for his fear that someone would come along and remove it, so he reluctantly unhooked the painting and took it back to his studio.

We now know where in the brain to find language, logic, and the impulse to create. We know how the brain signals the body to move, to fight and to flee, to love and to sleep, to write and to weep. We have pinpointed each of the senses—how and where we see, smell, and feel the petals of a single rose. But where in the brain's gelatinous gray matter is the "I" that perceives the rose's exquisiteness? Where is the "I" that experiences its beauty? Looking for the subjective self, for the soul, is like looking for the essence of Beethoven's Fifth Symphony in the sheet music. The score is merely a combination of the notes played by the violins, the cellos, the timpani. But where is the power, the passion, of this thing called Beethoven's Fifth? As Emily Dickinson once wrote, ironically, "Split the lark—And you'll find the music." Except that you won't. You can't. The whole *is* more than the sum of its parts, and even the parts sometimes seem like mirages. Every idea, desire, memory, or emotion is nothing more than a leap of chemicals across a synaptic abyss—a chasm between two neurons ten times smaller than the width of an atom. The brain, Dickinson wrote in another poem, is "just the weight of God." It is incomprehensibly multiple, and inexplicably singular, and it is quite simply what makes us who we are.

Jon Sarkin would keep creating. He would sift through all the possibilities. And then he would paint them, write them, into life—line by line, color by color. The words and images would take him where they would. Sometimes into the past, always into his imagination. In his mind, he would keep returning to the desert, looking for the story, needing to find it and tell it, over and over, until he got it right.

He was Jon Sarkin. He was nobody.

He was traveling alone, with Kim, or with Tony.

It was 1988, or it was tonight.

We drive through the desert. We pass a sign advertising "Scenic mesa hot springs." The hot springs consist of foul pools of sulphurous waters. These will be hard miles. Tony and I have not slept since Topeka and we are both a jangle of bad craziness. It is 3 a.m., the sky is as still as a dark tornado. I flip on the radio. Nothing but static and farm reports.

It was today, or it was yesterday.

Signs point out motels and roadside attractions. "Mesa Motel 4 miles," "Canyon Falls next exit." It's night, early morning actually, and the sky is as black as a tornado. The only light in this bleak god-forsaken wasteland is an occasional oncoming car. We haven't slept since Chicago and my handle on life is broken.

There are pools of perception that lie untouched in most human be-ings, like the water beneath a frozen river that churns and swirls but remains invisible. Sarkin's stroke had punched a hole through the ice. Set loose from the constraints of normal reasoning, he had plunged into a riotous river of unmediated perception. His brain refocused on the random details of life, mixing memory and emotion, then distill-ing his experiences into words and images. He knew he was no longer the man he had been. His brain was a broken mirror, its pieces reflect-ing all the different parts—the husband, the child, the father, the art-ist, the writer, the obsessive. He had become, like one of his favorite poets Wallace Stevens once wrote, the sum of all "human shadows, bright as glass." He would keep trying to stitch it back together, even as he knew he never could. Every picture, every poem, every paint-ing was still just a rough draft. He would always have to create new stories, sift through more possibilities, line by line, color by color. The imperfect was now his paradise, and his art, like his life, lay in flawed words and stubborn sounds.

BIBLIOGRAPHY

Abrams, Michael. "Sight Unseen." *Discover*, June 2002.

Abse, Dannie. *A Poet in the Family.* Robson Books, 1984.

———. *White Coat, Purple Coat.* Persea, 1991.

Aharon-Peretz, et al. "Characterization of Empathy Deficits Following Prefrontal Brain Damage: The Role of the Right Ventromedial Prefrontal Cortex." *Journal of Cognitive Neuroscience*, vol. 15, no. 3, 2003.

Andreasen, Nancy. *The Creative Brain: The Science of Genius.* Penguin, 2005.

Anonymous. "Great Teachers of Surgery in the Past: Lambert Charles Rogers." *British Journal of Surgery*, vol. 51, no. 11, Nov. 1964.

Balaban, Evan. "Brain Switching: Evolutionary Behavioral Changes in the Context of Individual Brain Development." *International Journal of Developmental Biology*, vol. 49, 2005.

Ball, Philip. "Natural Talent." *New Scientist*, Oct. 29, 2005.

Balzac, Fred. "Sudden Emergence of Visual Creativity in Patients with Frontotemporal Dementia." *Neuropsychiatry Reviews*, June 2003.

Bargh, John, and Chartrand, Tanya. "The Unbearable Automoticity of Being." *American Psychologist*, July 1999.

Barker, Fred. "The American Crowbar Case and Nineteenth-Century Theories of Cerebral Localization." *AANS Bulletin*, Spring 1993.

Barrett, William. *Death of the Soul: From Descartes to the Computer.* Anchor Press/Doubleday, 1986.

Bennett, Drake. "Thinking Literally: The Surprising Ways That Metaphors Shape Your World." *Boston Globe*, Sept. 27, 2009.

Benton, Arthur, ed. "The Memoir of Marc Dax on Aphasia," in *Exploring the History of Neuropsychology: Selected Papers.* Oxford University Press, 2000.

Bermudez-Rattoni, Federico. *Neural Plasticity and Memory: From Genes to Brain Imaging.* CRC Press, 2007.

Berns, Gregory. "Neuroscience Sheds New Light on Creativity." *Fast Company*, Sept. 17, 2008.

Bigelow, H. J. "Dr. Harlow's Case of Recovery from the Passage of an Iron Bar Through the Head." *American Journal of the Medical Sciences*, vol. 19, 1850.

Blackburn, Simon. "The World in Your Head." *New Scientist*, Sept. 11, 2004.

Blakeslee, Sandra. "Cells That Read Minds." *New York Times*, Jan. 10, 2006.

Block, Ned, ed. *The Nature of Consciousness: Philosophical Debates.* MIT Press, 1997.

Bogousslavsky, Julien, and Boller, Francois. *Neurological Disorders in Famous Artists.* Karger, 2005.

Bolhuis, Johan, and Wynne, Clive. "Can Evolution Explain How Minds Work?" *Nature*, vol. 458, April 16, 2009.

Bosveld, Jane. "Soul Search." *Discover*, June 2007.

Boyd, Brian. *On the Origin of Stories: Evolution, Cognition and Fiction.* Belknap Press, 2009.

Breathnach, Caoimhghin, and Ward, Conor. "The Victorian Genius of Earlswood—A Review of the Case of James Henry Pullen." *Irish Journal of Psychiatry and Medicine*, vol. 22, no. 4, 2005.

Brink, T. L. "Idiot Savant with Unusual Mechanical Ability: An Organic Explanation." *American Journal of Psychiatry* 137:2, Feb. 1980.

Brockman, John. *The Third Culture: Beyond the Scientific Revolution.* Simon & Schuster, 1995.

Broks, Paul. "Exorcising the Ghost in the Machine." *Financial Review,* July 18, 2002.

———. *Into the Silent Land.* Atlantic Monthly Press, 2003.

———. "Mirror, Mirror on the Wall, Is There Anyone There at All?" *Times* (London), Sept. 20, 2005.

———. "The Mystery of Consciousness." *Issue*, vol. 133, April 2007.

Brown, Andrew. "How to Make a Soul." *Wired*, Feb. 2008.

Brown, Bob. "Victims of Brain Trauma Driven to Create." *ABC News*, Sept. 9, 2008.

Byron, Paula. "The Incurable Disease of Writing." *Harvard Medical Alumni Bulletin*, Autumn 2003.

Carroll, Joseph. *The Adaptive Function of Literature and the Other Arts.* National Humanities Center, 2006.

Carroll, Linda. "The Art of Therapy." *Neurology Now*, Nov./Dec. 2006.

———. "The Science of Art." *Neurology Now*, Nov./Dec. 2006.

Carter, Rita, and Frith, Christopher. *Mapping the Mind* (Phoenix, 2000).

Chatterjee, Anjan. "The Neuropsychology of Visual Art: Conferring Capacity." *International Review of Neurobiology*, vol. 74, 2006.

———. "The Neuropsychology of Visual Artistic Production." *Neuropsychologia*, vol. 42, 2004.

———. "Prospects for a Cognitive Neuroscience of Visual Aesthetics." *Bulletin of Psychology and the Arts*, vol. 4, 2003.

Clark, Andy. *Supersizing the Mind: Embodiment, Action and Cognitive Extension.* Oxford University Press, 2008.

Coghlan, Andy. "How Culture Made Your Modern Mind." *New Scientist,* May 14, 2008.

Colapinto, John. "Brain Games: The Marco Polo of Neuroscience." *New Yorker,* May 11, 2009.

Colburn, Dan. "Scientists Discover Brain's Adaptability." *Washington Post,* Sept. 28, 1999.

Corsello, Andrew. "Metamorphosis," *GQ,* Jan. 1997.

Counter, S. Allen. "Music Stirred Her Brain." *Boston Globe,* Mar. 29, 2005.

Cranny-Francis, Anne. "Sonic Assault to Massive Attack: Touch, Sound and Embodiment." *Scan,* vol. 5, no. 7, Dec. 2008.

Crick, Francis. *The Astonishing Hypothesis: The Scientific Search for the Soul.* Scribner, 1995.

Cromie, William. "Thought of Pain Changes the Brain." *Harvard University Gazette,* Feb. 16, 2004.

Crompton, Simon. "Painting? I Can't Turn It Off." TimesOnline (London), July 14, 2007.

Custance, Arthur. *The Mysterious Matter of Mind.* Zondervan, 1979.

Damasio, Hanna, et al. "The Return of Phineas Gage: The Skull of a Famous Patient Yields Clues About the Brain." *Science,* vol. 264, May 20, 1994.

Demaree, Glenda. "Hyperacusis." The Hyperacusis Network, www.hyeracusis.net.

De Waal, Frans. *The Age of Empathy.* Harmony Books, 2009.

Doidge, Norman. *The Brain That Changes Itself: Stories of Personal Triumph from the Frontiers of Brain Science.* Penguin, 2007.

Dolan, Brian. "Soul Searching: A Brief History of the Mind/Body Debate in the Neurosciences." *Neurosurgical Focus,* vol. 23, July 2007.

Dupree, Catherine. "Authorial Synapses." *Harvard,* Jan.-Feb. 2004.

Dutton, Denis. "The Pleasures of Fiction." *Philosophy and Literature,* vol. 28, 2004.

Eagleman, David. "10 Unsolved Mysteries of the Brain." *Discover,* Aug. 2007.

Edelman, Gerald. *Wider than the Sky: The Phenomenal Gift of Consciousness.* Yale University Press, 2004.

Eliot, T. S. *The Complete Poems and Plays of T. S. Eliot.* Faber & Faber, 2004.

Feinberg, Cara. "Old Brain, New Tricks." *Boston Globe,* Jan. 15, 2006.

Feinberg, Todd. *Altered Egos: How the Brain Creates the Self.* Oxford University Press, 2002.

Feinberg, Todd, and Keenan, Julian. *The Lost Self: Pathologies of the Brain and Identity.* Oxford University Press, 2005.

Flaherty, Alice. *The Midnight Disease: The Drive to Write, Writer's Block, and the Creative Brain.* Mariner Books, 2005.

Fodor, Jerry. *The Mind Doesn't Work That Way: The Scope and Limits of Computational Psychology.* MIT Press, 2001.

Fox, Douglas. "The Inner Savant." *Discover,* Feb. 1, 2002.

Friston, Karl. "The Prophetic Brain." *Seed,* July 20, 2009.

Frith, Chris. *Making Up the Mind: How the Brain Creates Our Mental World.* Blackwell, 2007.

Gardner, Howard. *Extraordinary Minds: Portraits of 4 Exceptional Individuals and an Examination of Our Own Extraordinariness.* Basic Books, 1997.

———. "The Philosophy-Science Continuum." *Chronicle of Higher Education,* March 9, 2001.

Gaufo, Gary, Thomas, Kirk, and Capecchi, Mario. "Hox3 Genes Coordinate Mechanisms of Genetic Suppression and Activation in the Generation of Branchial and Somatic Motor Neurons." *Development,* Nov. 1, 2003.

Gawande, Atul. "The Itch: Its Mysterious Power May Be a Clue to a New Theory About Brains and Bodies." *New Yorker,* June 30, 2008.

Gazzaniga, Michael. *Cognitive Neuroscience: The Biology of the Mind,* 3d ed. W. W. Norton, 2008.

———. *The Ethical Brain: The Science of Our Moral Dilemmas.* Harper Perennial, 2006.

———. *Human: The Science Behind What Makes Us Unique.* Ecco, 2008.

Gazzaniga, Michael, and LeDoux, Joseph. *The Integrated Mind.* Springer, 1978.

Goguen, Joseph, ed. "Art and the Brain." *Journal of Consciousness Studies,* vol. 6, 1999.

Gordon, N. "Unexpected Development of Artistic Talents." *Postgraduate Medical Journal,* vol. 81, 2005.

Grady, Denise. "The Vision Thing: Mainly in the Brain." *Discover,* June 1, 1993.

Greenberg, Neil. "The Beast at Play: The Neuroethology of Creativity," in *The Child's Right to Play: A Global Approach.* Praeger Press, 2004.

Grimm, S., et al. "Imbalance Between Right and Left Dorsolateral Prefrontal Cortex in Major Depression Is Linked to Negative Emotional Judgment." *Biological Psychiatry,* vol. 63, 2008.

Grist, Matt. "Surprise Yourself: The Social Brain." *RSA Journal,* July 15, 2009.

Grüter, Thomas, and Kraft, Ulrich. "Alien Friends." *Scientific American Mind,* April 2005.

Gültekin, S., et al. "Vascular Loops at the Cerebellopontine Angle: Is There a Correlation with Tinnitus?" *American Journal of Neuroradiology,* July 24, 2008.

Hansotia, Phiroze. "A Neurologist Looks at Mind and Brain." *Clinical Medicine & Research,* Oct. 2003.

Harlow, J. M. "Passage of an Iron Rod Through the Head." *Boston Medical and Surgical Journal,* 1848, rep. in *Journal of Neuropsychiatry and Clinical Neuroscience,* vol. 11, Feb. 1999.

———. "Recovery from the Passage of an Iron Bar Through the Head." *Publications of the Massachusetts Medical Society,* vol. 2, 1868.

Hasson, Uri, et al. "Neurocinematics: The Neuroscience of Film." *Projections,* vol. 2, no. 1, summer 2008.

Heller, Rick. "Science and Religion." Freeminds.org, vol. 16, Dec. 23, 2008.

Hofer, Petra, and Rockenhaus, Freddie (writers and producers). *Beautiful Minds—A Voyage into the Brain.* German documentary, 2003.

Huang, Gregory. "Is This a Unified Theory of the Brain?" *New Scientist*, May 28, 2008.

Hume, David. *Enquiries Concerning Human Understanding and Concerning the Principles of Morals,* 3d ed. Oxford University Press, 1975.

———. *A Treatise of Human Nature,* vol. 1. Oxford University Press, 2007.

Humphrey, Nicholas. *Seeing Red.* Belknap Press of Harvard University, 2006.

———. "The Thick Moment," in *The Third Culture: Beyond the Scientific Revolution,* by John Brockman. Simon & Schuster, 1995.

Iacoboni, Marco. *Mirroring People: The New Science of How We Connect with Others.* Farrar, Straus and Giroux, 2008.

James, William. *The Principles of Psychology.* Cosimo Classics, 2007.

———. *The Varieties of Religious Experience.* New American Library, 1958.

Jannetta, Peter, et al. "Technique of Microvascular Decompression." *Neurosurgical Focus,* vol. 18, April 2005.

Johnson, Mark. "Mind Incarnate: From Dewey to Damasio." *Daedalus,* June 22, 2006.

Kandel, Eric. "A Neuroscience Sampling." *Edge: The Third Culture,* March 5, 2007.

———. *Principles of Neural Science,* 4th ed. McGraw-Hill Medical, 2000.

Kaptchik, Ted. "Components of Placebo Effect: Randomised Controlled Trial in Patients with Irritable Bowel Syndrome." *British Medical Journal,* April 3, 2008.

Keenan, J., et al. "Self-Recognition and the Right Hemisphere." *Nature,* vol. 409, 2001.

Koch, Christof. "The Movie in Your Head." *Scientific American Mind,* Oct. 2005.

———. *The Quest for Consciousness: A Neurobiological Approach.* Roberts & Company Publishers, 2004.

Kraft, Ulrich. "Unleashing Creativity." *Scientific American Mind,* April 2005.

La Cerra, Peggy, and Bingham, Roger. *The Origins of Minds: Evolution, Uniqueness, and the New Science of the Self.* Harmony Books, 2002.

LeDoux, Joseph. *The Synaptic Self: How Our Brains Become Who We Are.* Penguin, 2003.

Lehrer, Jonah. "The Eureka Hunt." *New Yorker,* July 28, 2008.

Leiner, Henrietta, and Leinor, Alan. "The Treasure at the Bottom of the Brain." *New Horizons for Learning,* Sept. 1997.

Lieberman, Matthew, et al. "The Neural Correlates of Placebo Effect: A Disruption Account." *Neuroimage,* vol. 22, June 2004.

Livingstone, Margaret. "Light Vision." *Harvard Medical Alumni Bulletin,* Spring 2005.

Long, Kevin, et al. "Transferring an Inborn Auditory Perceptual Predisposition with Interspecies Brain Transplants." *Proceedings of the National Association of Science,* vol. 98, no. 10, May 8, 2001.

Lythgoe, Mark, et al. "Obsessive, Prolific Artistic Output Following Subarachnoid Hemorrhage." *Neurology*, vol. 64, 2005.

Macmillan, Malcolm. *An Odd Kind of Fame: Stories of Phineas Gage*. MIT Press, 2002.

MacNeilage, Peter, et al. "Origins of the Left & Right Brain." *Scientific American*, July 2009.

Malafouris, Lambros. "The Cognitive Basis of Material Engagement: Where Brain, Body and Culture Conflate." *Philosophical Transactions of the Royal Society*, vol. 363, 2008.

Malhotra, S. "Autism: An Experiment of Nature." *Journal of the Indian Association of Child and Adolescent Mental Health*, vol. 2, no. 1, 2006.

Marien, Peter, et al. "Cerebellar Neurocognition: A New Avenue." *Belgian Neurological Society*, 2001.

Martin, Douglas. "Obituary: Robert W. Shields." *New York Times*, Oct. 29, 2007.

Martin, Raymond, and Barresi, John. *The Rise and Fall of Soul and Self: An Intellectual History of Personal Identity*. Columbia University Press, 2006.

McCarty, Mary, and Wagner, Mike. "Flight of Angels," 6-part series. *Dayton Daily News*, October 2005.

McLaughlin, Mark, et al. "Microvascular Decompression of Cranial Nerves: Lessons Learned After 4,400 Operations." *American Association of Neurological Surgeons*, 1998.

Miller, Bruce. "Vital Signs: A Passion for Painting." *Discover*, Jan. 1, 1998.

Miller, Bruce, et al. "Portraits of Artists: Emergence of Visual Creativity in Dementia." *Archives of Neurology*, vol. 61, no. 6, June 2004.

Miller, Bruce, et al. "Unravelling Boléro: Progressive Aphasia, Transmodal Creativity and the Right Posterior Neocortex." *Brain*, vol. 131, 2008.

Millionaire, Tony. *The Art of Tony Millionaire*. Dark Horse, 2009.

Muangpaisan, Weerasak, et al. "The Alien Hand Syndrome: Report of a Case and Review of the Literature." *Journal of the Medical Association of Thailand*, vol. 88, 2005.

Myers, Nan. "Peter Jannetta Shows Them How It's Done." *Penn Medicine*, summer 2008.

Myers, P Z. "A Profound Sense of Time." *Seed*, Oct. 17, 2007.

Nash, Madeleine. "The Gift of Mimicry," in *The Brain: A User's Guide. Time* special issue, Jan. 29, 2007.

Noë, Alva. *Out of Our Heads: Why You Are Not Not Your Brain, and Other Lessons from the Biology of Consciousness*. Hill and Wang, 2009.

O'Connor, James. "Thomas Willis and the Background to *Cerebri Anatome*." *Journal of the Royal Society of Medicine*, vol. 96, 2003.

Oftedal, Gunnhild, et al. "Mobile Phone Headache: A Double-Blind, Sham-controlled Provocation Study." *Cephalagia*, vol. 27, no. 5, 2007.

Okamura, Takehiko, et al. "A Clinical Study of Hypergraphia in Epilepsy." *Journal of Neurology, Neurosurgery, and Psychiatry*, vol. 56, 1993.

Olson, Charles. *The Maximus Poems*. University of California Press, 1985.

Owen, Adrian, et al. "Detecting Awareness in the Vegetative State." *Science*, vol. 313, no. 5792, Sept. 8, 2006.

Parkin, Alan. *Explorations in Neuropsychiatry*. Blackwell, 1996.

Penfield, Wilder. *Mystery of the Mind: A Critical Study of Consciousness and the Human Brain*. Princeton University Press, 1978.

Phillips, Helen. "Looking for Inspiration." *New Scientist*, Oct. 29, 2005.

Pilcher, Helen. "Breaking the Voodoo Spell." *New Scientist*, May 13, 2009.

Pinker, Steven. *The Blank Slate: The Modern Denial of Human Nature*. Penguin, 2002.

———. *How the Mind Works*. W. W. Norton & Co., 1997.

Plot, Robert. "*The Natural History of Oxfordshire: Being an Essay Towards the Natural History of England*. Printed by Leon. Lichfield, for Charles Brome and John Nicholson, London, 1705.

Popper, Karl, and Eccles, John. *The Self and Its Brain: An Argument for Interactionism*. Routledge, 1984.

Raines, John. "Earth Vigil: Darwin, Death and Hope." *Cross Currents*, vol. 44, summer 1994.

Ramachandran, Vilayanur. *A Brief Tour of Human Consciousness: From Imposter Poodles to Purple Numbers*. Pi Press, 2004.

———. "Mirror Neurons and the Brain in the Vat." *Edge: The Third Culture*, Jan. 10, 2006.

———. "The Neurology of Self-Awareness." *Edge: The Third Culture*, 2007.

———. "Self Awareness: The Last Frontier." *Edge: The Third Culture*, Jan. 1, 2009.

Ramachandran, Vilayanur, and Blakeslee, Sandra. *Phantoms in the Brain: Probing the Mysteries of the Human Mind*. Fourth Estate Ltd., 1999.

Ramachandran, Vilayanur, and Oberman, Lindsay, "The Stimulating Social Mind: The Role of the Mirror Neuron System and Simulation in the Social and Communicative Deficits of Autism Spectrum Disorders." *Psychological Bulletin*, vol. 133, no. 2, 2009.

Rapoport, Mark, et al. "The Role of the Cerebellum in Cognition and Behavior." *Journal of Neuropsychiatry and Clinical Neuroscience*, vol. 12, no. 2, spring 2000.

Ratey, John. *A User's Guide to the Brain: Perception, Attention and the Four Theaters of the Brain*. Vintage Books, 2001.

Ratiu, P., Talos, I. F., Haker, S., Lieberman, D., Everett P. "The Tale of Phineas Gage." *Journal of Neurotrauma*, 2004.

Reltan, Albert. "Specialist's Factual Report of Investigation." National Transportation Safety Board, Vehicle Recorders Division, Nov. 15, 2004.

Restak, Richard. *The Naked Brain: How the Emerging Neurosociety Is Changing How We Live, Work and Love*. Harmony Books, 2006.

Rizzolatti, Giacomo, and Fabbri-Destro, Maddalena. "Mirror Neurons and Mirror Systems in Monkeys and Humans." *International Union of Physiological Science*, 2008.

Rogers, Lesley. "Seeking the Right Answers About Right Brain-Left Brain." *Cerebrum*, vol. 5, no. 4, Fall 2003.

Rosenfield, Israel, and Ziff, Edward. "How the Mind Works: Revelations." *New York Review of Books*, June 26, 2008.

Sapolsky, Robert. "Ego Boundaries, or the Fit of My Father's Shirt." *Discover*, Nov. 1995.

Schiff, Nicholas, and Fins, Joseph. "Hope for 'Comatose' Patients." The Dana Forum on Brain Science, 2007.

Schmahmann, Jeremy. "Disorders of the Cerebellum." *Journal of Neuropsychiatry and Clinical Neurosciences*, Aug. 2004.

Schmitz, Taylor, et al. "Metacognitive Evaluation, Self-Relevance, and the Right Prefrontal Cortex." *Neuroimage*, vol. 22, June 2004.

Schwartz, Jonathan. "'The Quarries Are Silent': Notes on a Scandinavian Community in New England." *American Studies in Scandinavia*, vol. 20, 1988.

Schwitzgebel, Eric, and Gordon, Michael. "How Well Do We Know Our Own Conscious Experience? The Case of Human Echolocation." *Philosophical Topics*, vol. 28, no. 2, Fall 2000.

Searle, John. *Minds, Brains and Science*. Harvard University Press, 1984.

Serrell, Orlando. The Orlando Sorrell official website, www.orlandosorrell.com.

Shaffer, Jerome. "Personal Identity: The Implications of Brain Bisection and Brain Transplants." *Journal of Medicine and Philosophy*, vol. 2, no. 2, 1977.

Shelton, Mark. *Working in a Very Small Place: The Making of a Neurosurgeon*. Vintage Books, 1990.

Shermer, Michael. "Patternicity." *Scientific American*, Nov. 25, 2008.

Shreeve, James. "What Happened to Phineas." *Discover*, Jan. 1995.

Silberman, Steve. "The Placebo Problem." *Wired*, Sept. 2009.

Snyder, Allan, et al. "Savant-Like Skills Exposed in Normal People by Suppressing the Left Fronto-Temporal Lobe." *Journals of Integrative Neuroscience*, vol. 2, no. 2, 2003.

Spears, Tom. "Search the Brain to Find the Soul, Surgeon Says." *Ottawa Citizen*, Nov. 7, 2003.

Speyrer, John. "The Role of the Temporal Cortex in Certain Psychical Phenomena." *Journal of Mental Science*, July 1955.

Squire, Larry, ed. *History of Neuroscience in Autobiography*. Society for Neuroscience, 2006.

Stein, Kathleen. *The Genius Engine: Where Memory, Reason, Passion, Violence and Creativity Intersect in the Human Brain*. John Wiley & Sons, 2007.

Stevens, Wallace. *The Collected Poems*. Knopf, 1954.

Sutton, John. "Memory, Embodied Cognition, and the Extended Mind." *Philosophical Psychology*, vol. 19, no. 3, June 2006.

Thompson, Richard F. *Memory: The Key to Consciousness*. Princeton University Press, 2007.

Thomson, Helen. "Empathy Overkill." *New Scientist*, March 13, 2010.

Tolson, Jay. "Is There Room for the Soul?" *U.S. News & World Report*, Oct. 23, 2006.

Tooby, John, and Cosmides, Leda. "Does Beauty Build Adaptive Minds?" *SubStance*, no. 94/95, 2001.

Treffert, Darold. "Savant Syndrome: 2000 and Beyond." Wisconsin Medical Society, 2000.

Treffert, Darold, and Wallace, Gregory. "Islands of Genius." *Scientific American*, June 2002.

Vallance, Aaron. "Something Out of Nothing: The Placebo Effect." *Advances in Psychiatric Treatment*, vol. 12, 2006.

Van Baaren, Rick, et al. "Mimicry for Money: Behavioral Consequences of Imitation." *Journal of Experimental Social Psychology*, vol. 39, 2003.

Van Helden, Albert, and Burr, Elizabeth. "The Galileo Project." Rice University, http://galileo.rice.edu/.

Volkan, Vamik. "Obituary: David Wilfred Abse, M.D." *Group Analysis*, vol. 40, no. 4, 2007.

Willingham, Elizabeth. "The History of Tinnitus." Baylor College of Medicine, 2004.

Zeki, Semir. "Artistic Creativity and the Brain." *Science*, vol. 293, no. 5527, July 6, 2001.

Zimmer, Carl. "How Google Is Making Us Smarter." *Discover*, Feb. 2009.

———. "The Neurobiology of the Self." *Scientific American*, Nov. 2005.

———. *Soul Made Flesh: The Discovery of the Brain—and How It Changed the World*. Free Press, 2004.

ACKNOWLEDGMENTS

This book was a labor of love because of the two people at its heart—Jon and Kim. I cannot thank them enough for their time, their honesty, and their courage. And to their three children, Curtis, Robin, and Caroline: thank you for your indulgence. Many thanks as well to Jon's mother Elaine Zheutlin and her husband Bill, to Jon's sister Jane Sarkin and her husband Martin O'Connor, and to the late Richard Sarkin and his widow, Marcia, as well as to Jon's many friends.

To my super-agent, Wendy Strothman, and my super-hero editor, Emily Loose, I extend my humble thanks as well.

To my mother and father, Grace and Dave Nutt; my brother, Ty, my sisters Eva Nies, Cora Chemidlin, and Katie Barry; my brothers-in-law Rob Nies, David Chemidlin, and Pat Barry and sister-in-law Janie Nutt, and to my twelve amazing nieces and nephews: Brendan, Evan, and Cullen Nutt; Rachel, Maddie, and Jordan Nies; Conor, Blair, Grant, and Reid Chemidlin; Bridget and Patrick Barry and all my aunts, uncles, and cousins, especially Bess, and my godparents Betts and Bills, Betty and Bill Horn—I love you all more than life itself.

To my former *Star-Ledger* editor and writer extraordinaire Rosemary Parrillo, and to my current editor, the gifted Mr. Poetry Pants, David Tucker, thank you for your wisdom, guidance, patience, and support. To columnist Mark Di Ionno and former editor Susan Olds, who were responsible for hiring me at *The Star-Ledger* all those years ago, and former editor Deborah Jerome, who was an early supporter of my work: you all helped to make me into the writer I am today.

To my mentor and colleague at the Columbia University Graduate School of Journalism, Sandy Padwe—you are, and always will be, the best—and my great teachers there, especially Sam Freedman, Helen Benedict, and the incomparable Judith Crist. To Jane "Bambi" Wulf, thank you for my first job in journalism.

This book surely would never have been written without the former editor of *The Star-Ledger*, Jim Willse, who started me on my newspaper career in 1997 and saw the potential for Jon Sarkin's story when he agreed to publish "The Accidental Artist" in 2008: Chief, thanks will never be enough. My deep gratitude to Andre Malok and Jennifer Brown, who toiled away on that project, and to the current editor of *The Star Ledger*, Kevin Whitmer.

And to my dearest friends, your love has sustained me through many a trying time: Robin Gaby Fisher, the journalist who inspired me most; the brilliant Linda Kost and Wylie Willson; the phenomenal Dr. Jane (Maxwell) McInerny, whose strength of character and great good humor I admire beyond words; the indomitable Eileen Marr; and my oldest college buddy, Kathleen (BJ) Howes, and all my Nieman friends. To Ray Parrillo, the incomparable Philly sportswriter, thanks for the meals at the Office, and your insight on all things athletic. For all the laughs, thank you to Chris Mitchell and Angela of the Consonants. For all the extra love, thank you to "Mother Joan" Barradale.

Finally, to my first literary hero in life, my tenth-grade English teacher, the late Mrs. Janet B. Kollmar, and my first literary inspiration, my fifth-grade teacher, Marjorie Looby—I hope I've made you proud.

INDEX

ABOUT THE AUTHOR

Amy Ellis Nutt is a journalist and writer for *The Star-Ledger* (NJ). She was a finalist for the 2009 Pulitzer Prize in newspaper feature writing for her series about Jon Sarkin and a Nieman Fellow in Journalism at Harvard. She has won numerous national awards, including the American Association for the Advancement of Science's Pinnacle of Excellence Award (2004) and the American Society of Newspaper Editors' first place award for Non-deadline Writing for a series of five science stories that were published in *The Best Newspaper Writing of 2003*. She lives in Watchung, New Jersey.